Integrated Nutrient Management for Sustainable Agriculture

The Author

Dr Subhash Chand obtained M.Sc. (Ag.) in Soil Science from Swami Keshwanand Rajasthan Agricultural University, Bikaner and Ph.D. in Soil Science from MPUA&T, udaipur. Dr Chand is currently working as an Associate Professor at Faculty of Agriculture Wadura Campus, Sopore of Sher-e-Kashmir University of Agricultural Sciences and Technology of Kashmir Shalimar, Srinagar. He has published several research papers in journals of repute. Dr Subhash Chand is recipient of PG course work seminar award and Somani's book award of ISSS, Udaipur chapter and PG Merit Scholarship of the University. He is a life member of several scientific societies.

Integrated Nutrient Management for Sustainable Agriculture

– *Author* –

Dr Subhash Chand

Associate Professor

Division of Soil Science

Sher-e-Kashmir University of Agricultural Sciences and Technology of Kashmir, Shalimar Campus,

Srinagar – 190 025, J&K (India)

2019

Daya Publishing House®

A Division of

Astral International Pvt. Ltd.

New Delhi – 110 002

Published by　　　　　：　**Daya Publishing House®**
　　　　　　　　　　　　　A Division of
　　　　　　　　　　　　　Astral International Pvt. Ltd.
　　　　　　　　　　　　　– ISO 9001:2015 Certified Company –
　　　　　　　　　　　　　4736/23, Ansari Road, Darya Ganj
　　　　　　　　　　　　　New Delhi-110 002
　　　　　　　　　　　　　Ph. 011-43549197, 23278134
　　　　　　　　　　　　　E-mail: info@astralint.com
　　　　　　　　　　　　　Website: www.astralint.com

Digitally Printed at　　：　**Replika Press Pvt. Ltd.**

Preface

Soil is a precious natural resource equally as important as water and air. The proper use of soil greatly determines the capability of a life-support system and the socio-economic development of a country. The agriculture era has been changed from resource degrading to resource conserving technologies and practices which will enable help for increasing crop productivity besides maintaining soil health for future generations. Green revolution besides achieving food security imposes several threats like deterioration of the soil organic carbon stock (SOC), decreasing factor productivity, imbalances in NPK and micronutrient uses and disparity in fertilizer consumptions *etc.* The Integrated Nutrient Management (INM) provides an excellent opportunity not only for sustaining soil but enhancing crop productivity also. The INM is the maintenance or adjustment of soil fertility and plant nutrient supply to an optimum level for sustaining the desired crop production through optimization of the benefits from all possible sources of plant nutrients in an integrated manner. The information on INM practices is wider and scattered, hence an effort have been made for systematic and crop wise compilation of various INM packages along with theoretical explanation for better understanding of the subject.

The book is the result of the assimilation of knowledge, research work and experimentation and research papers of numerous scientists and researchers which have been appropriately compiled. The author consulted a variety of research papers, books, journals, magazines, proceedings as cited in the heading "Literature Consulted" and express his gratitude to the writers of the same.

The present book contains 200 INM options covering almost all the crops rising in a variety of soils in various states of the country. The book suggests a clear-cut INM options to farmers, scientists, research workers for the best management of all plant nutrient resources for sustaining crop yields and maintaining inherent capacity of a soil to supply plant nutrients in a balanced manner. The information given in the appendices will be fruitful for students, researcher and scientist for better understanding of the subject.

Subhash Chand

Contents

Chapter 1
Properties of Soil that Determines Soil Fertility

God is Bread to Hungry People

— Mahatma Gandhi

Agriculture is facing several critical issues like low fertilizer use efficiencies, decreasing factor productivity, low soil organic carbon (SOC) stock, mismatch between nutrient removal and addition to the soil (10-12 mt. of nutrient gap), disparities in the NPK ratio and water scarcity. The whole scenario of agriculture is at a junction where one has to rethink and reform the agricultural packages and practices to fulfill the dreams of million peoples of the country. All these challenges are resultant of green revolution by which we have attained self sufficiency in food grains production. Now in the next 50 years, the challenges for agriculture will not only be to meet the food needs of the worlds expanding population but also undertake it in a manner that is sustainable and profitable for future generation. The fertilizers are very important sources of plant nutrients and played a prominent role in increasing food grain production of the country. About fifty per cent increase in the food grain production in post green revolution era is attributed to the use of fertilizers.

There are several properties, which are responsible for the soil fertility and directly or indirectly play an important role in sustaining crop yields.They are:

Soil pH

☆ is a measure of the soil acidity or alkalinity of a soil.

☆ Measures the negative logarithm of the hydrogen ion activity of the soil solution.

☆ The scale ranges from 0 to 14 with pH 7.0 as the neutral point.

Soil Colour

☆ Reddish to brownish colour shows well-drained conditions. Degree of yellowness and motting shows poor drainage. Gray to dark colour indicates the presence of organic material.

☆ Soil colour is a result of the light reflected from the soil. It may be divided in three part (1) Hue: denotes the dominant spectral of colour (red, yellow, blue and green (2) Value: denotes the lightness or darkness of the colour (3) Chroma: represents the purity of the colour.

Bulk Density

☆ The bulk density is defined as the mass (weight) per unit volume of a dry soil including pore space.

☆ A compact soil has a higher value while an organic soil has a lower value.

☆ Generally in normal soils BD ranges from 1.0-1.65 Mg/m^3.

Soil Depth

☆ Refers to the depth to which plant roots can penetrate the soil and is the distance between the lowest and the upper most horizons of the soil.

Field Capacity (Soil Water)

☆ Refers to the moisture content of a soil after the loss of gravitational water.

☆ At this point, water is held in soil micropores, which is available to plant roots, until the water content down to a lower value and named as capillary water.

☆ This lower value is referred to as the permanent wilting point (-0.1 to –15 bars)

☆ The amount of water available between field capacity and permanent wilting point referred to as available water.

☆ This value is important in determining irrigation intervals.

Water Holding Capacity

☆ Sandy soils have a low capacity than clayey soils. This property is important for irrigation.

Soil Acidity

It is caused by many factors such as:

☆ Excessive rain which leaches basic cations (Ca, Mg, K)

☆ Use of nitrogenous fertilizer like urea, ammonium sulphate *etc.*

☆ Oxidation of iron pyrite containing minerals.

Correction of Soil Acidity

☆ Soil acidity is corrected by liming, soils which have a pH value of less than 5.5 could be limed.

Cation Exchange Capacity (CEC)

☆ The power to retain cations at the surface of soil colloids is referred to as the cation exchange capacity.

☆ Soil colloids of clay and organic matter have this property due to the presence of negative charges at the surface.

☆ CEC is defined as the sum of cations held by a kilogram of soil. It is expressed in Cmol/kg soil.

☆ Clays like kaolinite have low CEC and the CEC is pH dependent Organic matter has a large CEC but it too is pH dependent.

☆ Montomorillinitic clay has high CEC due to the negative charges developed through loss of cations during formation of these clays.

☆ Practices like fertilization, lining, irrigation and addition of organic manures can increase the exchangeable cations.

Water Logging

☆ State where soil becomes saturated with water thin due depth of the root zone for a period that affects yield and quality of crops.

☆ It may be of various reasons-over irrigation, seepage from canal, inadequate drainage or presence of a hard pan.

☆ Land under water logged conditions can be neither used for agriculture purpose nor for human settlements.

☆ This can be tackled by adopting scientific norms for amount or irrigation, checking seepage from canals by proper lining and providing adequate drainage through field channels.

Deterioration of Soil Structure

☆ It is a physical degradation of soil by heavy machinery used to cultivate the soil.

☆ This problem may be seen in soils having intermediate texture and very low organic matter content.

☆ It leads loss of stability of aggregates, scaling compaction and crusting resulting in poorer infiltration rate, high runoff and erosion.

Humus

It the portion of the organic material that has been transformed into a relatively stable form by soil microorganisms (organic colloids).

Salinity and Alkalinity

- ☆ Occurs in arid and semi arid regions, where precipitation is insufficient to meet evapotransporation needs of plants, when salts move up to the surface.
- ☆ Salt affected soils occur within irrigated lands.
- ☆ Salts are added through irrigation water, through over-irrigation and salts accumulate in poorly drained areas.
- ☆ Salt content in soils is measured in terms of electrical conductivity (EC).
- ☆ A saline soil has a pH of less than 8.5 but soils are well flocculated. The EC is more than 4 deci seimens per meter ($4dSm^{-1}$)
- ☆ Soils which are saline cause problems to crops during dry weather. Loss of crop yield from poor growth is common.
- ☆ Alkaline soils have high pH and a high concentration of sodium in them.
- ☆ Alkaline soils are deflocculated, drainage is poor and growing plants is difficult due to high pH and higher content of sodium in the soil.
- ☆ Draining, flushing after ploughing and addition of organic matter and gypsum could correct these problems.

Soil Organic Matter (SOM)

- ☆ Consists of living organisms, dead plant and animal residues.
- ☆ Is the most chemically active portion of the soil.
- ☆ Is a reservoir for various essential elements.
- ☆ Contributes to CEC.
- ☆ Promotes good soil structure
- ☆ Buffers soil pH
- ☆ Promotes good air and water relations in plants.

Chemical Nature of Soil Organic Matter

According to the majority of research workers in this field, soil organic matter can be classified into two groups:

Non-humic Substances

Non- humic substances consist of plant and animal substances in various stages of decomposition, the decomposition products of plant and animal substances, the organic compounds excreted by soil micro-organisms, and the bacterial plasma. These non-humic substances include all types of organic substances, namely proteins, lignin's, carbohydrates, fat, waxes, tannins, microbial, pigments and low molecular weight compounds.

Humic Substances

A major fraction of soil organic matter consists of high molecular weight substances which are comparatively resistant to decomposition and acidic in nature.

They do not belong to any class of organic compounds. In view of regularity of their occurrence in soils and certain special properties, it is very probable that they are novel polyneric substances, produced under unique conditions prevalent in soil. These humic substances can be divided into humic acid, fulvic acid and humin fractions. These fractions differ from each other in their chain length and consequently in their molecular weight.

Functions of Organic Matter in Soil

☆ Supplies N, P and S to plant growth

☆ Increases CEC in a soil

☆ Have large surface area and has high CEC

☆ Holds up to twenty times their weight of water

☆ Holds cations and anions and releases them slowly

☆ The ratio of C: N: P: S is 10: 1: 0.5: 0.1

☆ Effects the breakdown of pesticides and weedicides.'

☆ Chelates micronutrients such as Zn, Mn and Cu making them more available to plant roots

☆ Buffers extreme acidity, salinity and alkalinity.

Fate of Organic Material Added to Soils

☆ Undergoes decomposition (only bio-degradable materials) through soil macro and micro fauna and flora

☆ Final degradation is made by soil microorganisms.

☆ The final product is a humus type material which has a C: N ratio around 9.1-12.1.

☆ All complex carbon products are converted into simple compounds.

1.1 Concept of INM

The continuous and imbalance use of fertilizers is adversely affecting the sustainability of agricultural production besides causing environmental pollution. The major issue for the sustainable agricultural production will be management of soil organic carbon and rational use of organic inputs such as animal manure, crop residues, green manure, sewage sludge and wastes known as integrated plant nutrient resource management. However, since organic manure cannot meet the total nutrient needs of modern agriculture, hence integrated use of nutrients from fertilizers and organic sources will be the need of the time. The possible reasons for the apparent decline in returns from the increased fertilizer applications include.

☆ Imbalanced N, P and K application and the latter two are being applied in too low amounts in some states.

☆ Deficiencies of secondary and micronutrients are appearing with increasing frequency.

☆ More fertilizers are being used on soils inherently poorer in fertility and/ or uncertain water supply as in dryland areas.

☆ The increased intensity of cropping together with changes in crop sequences *e.g.* cereal-cereal rotations in place of cereal-legume rotations.

☆ There may be an over-all decrease in soil organic matter status of soils.

☆ Built up of certain diseases, pests and weeds under intensive system of cropping.

☆ Use of fertilizers nutrients in high amounts in some intensively cultivated areas may lead to serious problems of deterioration of the soil quality as a result.

To overcome the negative effects of application of plant nutrients, both at low or imbalanced and high levels of input can be avoided by excellent management. Balanced fertilization supplemented with organic nutrient sources help in overcoming the hazards of nutrient depletion and of mining soil fertility. Integrated Nutrient Management (INM) provides excellent opportunities to overcome all the imbalances besides sustaining soil health and enhancing crop production. The concept of INM is the maintenance or adjustment of soil fertility and of plant nutrient supply to an optimum level for sustaining the desired crop production through optimization of the benefits from all possible sources of plant nutrients in an integrated manner. The INM as defined by Harmsen, here differs from the conventional nutrient management by more explicit considering nutrient from different sources, notably organic materials, nutrients carried over from previous cropping seasons, the dynamics, transformations and interactions of nutrients in soil, interaction between nutrients, their availability in the rooting zone and during growing season in relation to the nutrient demand by the crop. In addition it integrates the objectives of production, ecology, environmental and is an important part of any sustainable agricultural production system. The objectives of INM may be classified as short objectives and broad objectives are:

The Shorter Objectives of INM are:

☆ To maintain or enhance soil productivity through a balanced use of mineral fertilizers combined with organic and biological sources of plant nutrients.

☆ To improve the stock of plant nutrients in the soil.

☆ To improve the efficiency of plant nutrients, thus limiting losses to the environment.

The Broader Objectives of the INM are:

☆ To increase the availability of nutrients from all sources in the soil during the growing season.

☆ To match the demand of nutrients by the crop and supply of nutrients from all sources through the labile soil nutrient pool, both in space (the rooting zone) and time (the growing season).

☆ To optimize the functioning of the soil biosphere with respect to specified functions, such as the decomposition of organic matter (mineralization), the control of the pathogenic organisms by their natural enemies (predators), the biological formation of soil structure (aggregates, biopores), the decomposition of phytotoxic compounds *etc.*

☆ To minimize the losses of nutrients to the environment *e.g.*, through ammonia volatization and denitrification in the case of nitrogen, surface run-off and leaching beyond the rooting zone.

The INM is a prescription for excellent soil health. Increasing sustainability concerns and shortage of fertilizers require a system of nutrient management that make optimum use of nutrients from soil resources. Efficient recycling and use of organic as carrier of nutrients, synergistic and antagonistic effect of nutrients on plant nutrition in cropping system for purpose of developing appropriate fertilizer plans. In this context INM is necessary and we need to put it in place. The IPNS system is a "System Approach" to farming, possibly it should have four major components *i.e.* (a) on-site nutrient resource generation (b) mobilisation of off-site nutrient resources, (c) resource integration and (d) resource management. There is a direct relationship of soil health with different attributes of soil and management options for sustainable agriculture (Figure 1.1).

In recent years a lot of emphasis has given for integrated use of inorganic fertilizers and organic manure have become important for higher agricultural production. No single source of plant nutrients, be it chemical fertilizers, organic manure, crop residues, green manure or even-biofertilizers can meet the entire nutrient needs of crops in modern agriculture.

Issues Soil degradation	Strategy Alternative agriculture	Goal Sustainable agriculture
x Soil erosion x Leaching x Nutrient run-off x Waterlogging x Desertification x Acidification x Compaction x Crusting x Organic matter loss x Salinisation x Nutrient depletion x Soil pollution x Toxicant accumulation	x Skilled management x Crop rotation x Soil and water conservation x Conservation tillage x **Integrated nutrient management** x Integrated pest management x Crop/livestock/fisheries systems	x Productive/profitable x Energy conserving x Environmentally sound x Economically viable x Conserved natural resources x Improved health/food quality/safety. x Socially acceptable x Ecologically balanced

Soil Quality

Figure 1.1: Basic Relationship of Soil Quality with different Attributes of Soil and Management Options for Sustainable Agriculture.

1.2 Components of INM

1.2.1 Soil Reserves

Soils considered as a mother for nurturing plants and supply all the 16 essential and 4 beneficial plant nutrients. Nutrients are mostly found in organic and/or fixed mineral form. Plants can meet much of their nutritional requirement form this source, if managed properly, mainly through mineralization of soil organic matter. But due to continuous and intensive cultivation, the nutrient supplying capacity of soils has been decreased considerably. Therefore under any intensive agriculture system, special emphasis should be given to raising soil organic matter (SOM) to maintain soil nutrient and to reduce soil degradation. To enhance soil nutrient supply it is necessary to adopt appropriate soil management practices, such as improvement of soil physical conditions and addition of appropriate quantities of nutrients including micronutrients through mineral fertilizer, organic and biological sources.

1.2.2 Use of Mineral Fertilizers

Fertilizer are industrially and chemically manufactured source of plant nutrients. Various types and grades of fertilizer are available throughout the country for supplying major nutrients such as N, P, K and micronutrients. The fertilizer use levels differ widely between various soils and nutrient use is mostly imbalanced, favouring excess use of nitrogen. Balanced fertilization is known to improve fertilizer use efficiency (FUE) and at the same time profitability for the farmer. Using ever higher rates of nitrogen (urea mostly) alone with improved better varieties, the resulting higher yields also remove ever larger amounts of soil nutrients if not replenished and the FUE declines further resulting in stagnating and even declining yields. This leads to the paradox situation where statistics report the continuing increase in fertilizer use but the expected crop production increases are not taking place it is also known as yield plateau.

Apart form N, P, K and S and micronutrients such as zinc (Zn), iron (Fe) manganese (Mn) and boron (B) have also gained in importance in recent years. The secondary nutrient sulphur has become deficient over wide areas especially since the intensive use of high analysis fertilizer, urea, instead of sulphate of ammonia and TSP or DAP instead of single superphosphate or NPK compounds. The major effect of these and several other factors is the gradual decline in crop yields and fertilizer use efficiency.

1.2.2.1 Importance of Balanced Fertilizer Uses

The presence of adequate quantities of nutrients in forms accessible to plant roots in soil solution is essential to attain optimum crop growth. Environmental factors and soil plant nutrient content and its availability determine the capacity of a plant to absorb such nutrients. Soil fertility determines the capacity of a soil to supply nutrients. The capacity of a soil to supply nutrients depends on many factors of which moisture content plays a big role.

Role of Balanced Fertilization

☆ Plants have specific nutrient requirements based on crop species and cultivars. Some species have a higher requirement of certain nutrients. Cereal requires more N while legumes require more available phosphorus. Sugar cane, banana and tomato require more potassium.

☆ In practice soils have the capacity to supply calcium, magnesium and most micronutrients. But enhanced crop production limits the availability of these nutrients.

☆ Maintaining the balance between the crop nutrient requirement and the capacity of the soil to supply these nutrient is the basis of balanced fertilizer use. Supplementing the soil's capacity to supply the crop requirement has to be done by means of adding chemical fertilizer or organic manure to the soil.

☆ As understanding of the nutrient requirement is the most important aspect of balanced fertilizer use.

☆ Addition of zinc, magnesium and sulphur is necessary to sustain rice yields over 200 bu/ac.

☆ Balanced fertilization requires the use of both chemical fertilizer and organic manure to supply the requirement of major, minor and micronutrients.

☆ Practical experience shows that addition of sulphur, calcium, magnesium, zinc and boron are helpful to certain crops *e.g.* Tea and cocoa.

☆ Application of calcium at pod initiation for groundnut is beneficial.

☆ Application of molybdenum for heading of cauliflower is important.

Balanced Fertilization helps in

☆ Increases fertilizer use efficiency.

☆ Increases crop yield and quality.

☆ Improves nutrient value of the crop residues, which can be used as manure.

☆ Minimizes build up of high levels of residual nutrients, antagonisms and negative interactions.

☆ Can reduce pesticide requirements and produce healthier food.

☆ Reduces adverse effects of wasteful fertilizer application, *i.e.* eutrophication, groundwater pollution or air pollution.

Imbalanced Fertilizers Effect Crop Yield and Sustainability

☆ When the elements essential for efficient plant nutrition and economic production are low in availability or unbalanced, then chemical fertilizers and soil amendments are required to enhance crop yields.

☆ Incorrect fertilizer recommendation or unbalanced/inadequate availability of nutrients can lead to depletion of soil nutrient reserves and loss of plant nutrients supplied to excess.

☆ Lack of balance nutrients also encourages excessive uptake of these nutrients supplied in excess but with no benefit.

☆ Imbalanced fertilization is an uneconomic waste of scarce resources.

1.2.3 Organic Sources of Plant Nutrients

The sustainability of highly intensive cropping systems and the associated heavy mineral fertilizer use without organic manures is widely questioned. This has brought the almost forgotten farmyard manures (FYM) and composts back to the forefront. Regular applications of such organic manures not only supply all the various secondary and micronutrients, though in small quantities, but also improve the physical and biological properties of the soil. Furthermore, return to the farm is the best way to take care of the large amounts of animal waste produced in the commercial animal husbandry, pig and poultry farms, instead of dumping and degrading the environment.

1.2.3.1 Farm Yard Manure

Farm yard manure (FYM) is a very good source of plant nutrients however, traditionally does not receive the attention it deserves, as most farmers store their most valuable asset, their cattle/buffalo manure not in a systematic, but in a rather haphazard way. Storage of FYM in rural households in the country is in heaps exposed to sun, wing and rain, which accounts for substantial nutrient losses. FYM preparation needs improvement, adhering to strict and prompt coverage for shading and prevention of drying out by hot wind or washing out of nutrients with heavy rains (pollution hazard).

1.2.3.2 Composting

Compost is not a by- product of common farm activities, but has to be specially prepared for its own sake. The quality of the ripe compost after undergoing a heating process reaching at least 60°C to destroy harmful pathogens and weed seeds will depend on the raw material used and the attention given to proper composting by the farmer. The C: N ratio needs to be lowered to 20-15 and good quality compost should have no more than 30 per cent moisture, as no farmer wants to carry excess water to the field. Practically all sixteen plant nutrients are contained in compost, but in very small quantities. For phospho-compost preparation, it is suggested to mix 3 per cent of SSP with composting materials.

Composting is a labour and time–consuming process, which takes 3-6 months. To speed up the process in several countries, rapid composting technologies have been developed. With the use of *Trichoderma harzianium*, a fungal activator, decomposition of rice straw and other organic material with high C: N ratio, combined with animal manure is enhanced to 25 days. Commercially prepared composts marketed as organic fertilizer are available in the countries and used mainly for high quality vegetable production and horticultural use. Several private companies coming forward for manufacturing compost for farmers use. For details of composting technologies see appendices XVIII,XIX,XX.

1.2.3.3 Crop Residues

Crop residues are freely available sources of organic matter that available on –farm in large quantities are wheat and rice straws, maize stalks, and stovers of legumes and various pulses. Most of the crop residues are not collected for composting and nutrient recycling, but are used as animal feed (straws/stovers), burnt or left in the field for natural decomposition (fallen leaves and stubble).Crop residues in the long run also increase the OM content in the soil. Mulching with flesh straw or leaves is another good agronomic practice for conserving moisture, reducing soil erosion and for recycling of nutrients, if the partly decomposed mulching material is ploughed under for the following crop. Direct seeding of maize or soybean into mulch cover would be another good agronomic practice. Burning of straw which is still widely practiced by farmers as the fastest and least labour requiring method of disposal should be discouraged.

1.2.3.4 Green Manuring

Green manure crops such as *Sesbania aculeate* ploughed into the soil after 45-60 days may contribute about 30-40 kg/ha nitrogen for the following crop. However, due to limitations to grow a green manure crop, which occupies the land for several months and needs water and fertilizer, except N-just to plough it back into the soil. Wherever possible and feasible the growing of grain legumes such as groundnuts, soybeans, chickpeas, cowpeas or mungbean as cash crops, which maintain soil fertility and provide farmers with extra income and fodder from crop residues should be encouraged for improving soil physico-bio-chemical properties.

1.2.4 Biogas Slurries

Bio-gas plants in rural areas produce digested slurry as an end product, which could be applied directly in cultivated fields. Such slurry contains about 1.5-2.0 per cent nitrogen, 1.0 per cent phosphorus and a little over 1 per cent potassium. It is also a valuable source of micronutrients. Moreover, due to the heated digestion processes, biogas slurry is virtually free from weed seeds and pathogens.

1.2.5 Industrial Waste Materials

Most industrial waste materials as are valuable resources and should be properly managed and utilized. The large number of sugar cane processing factories in the region produces substantial quantities of organic by- products such as biogases, pith and press-mud. Even though some of the biogases and came residues are burnt as fuel in the sugar industry. So far only a small portion is mixed with press-mud, composted and recycled as organic fertilizer.

Agro-industries, such as fruit and vegetable processing, cotton ginneries, oil mills, breweries and distilleries, also produce large quantities of organic waste materials which need to be properly managed and utilized for nutrient recycling instead of dumping and polluting the environment.

1.2.6 City Refuse (Garbage, Sewage Sludge)

Increasing population and even faster growth of urban population will consequently lead to increasing amounts of urban waste, which would create enormous disposal problems if not properly recycled as a source of crop nutrients. Processed, composted solid organic wastes and sewage sludge provide both organic matter and valuable plant nutrients to crops. The transport from urban composting plants to the farming areas constitutes a major part of the cost of processed organic wastes for farmers. Marketing studies and advertising campaigns, attractive comparative prices together with a subsidy scheme to encourage the large-scale acceptance by farmers of urban compost should be considered. Subsidies, grants and credit should concentrate on transport and handling cost of such bulky products, which could nevertheless result in considerable saving in mineral fertilizer, for the farmers. As a rule of thumb the price per kg of nutrient in composted city refuse for the farmer should be at par or not considerably higher than the cost per kg nutrient in commonly used mineral fertilizer. The other not so easily quantifiable benefits of using organic fertilizer materials, such as increasing SOM, better water holding capacity, and better soil health, are to be accounted for by the cost of extra labour for spreading and incorporation in the field.

1.2.7 Enriched City Compost

City compost produced at mechanical composting plants throughout the country is generally low in plant nutrients and therefore its acceptability by farmers has been limited. To improve the quality and nutrient content of city compost low-grade rock phosphate and phosphate solubilising *azotobacter spp.* and the nitrogen fixing bacteria, such as *azotobactor spp.* or *pseudomonas spp.* are being used as inoculants. Microbial inoculation and application of 1 to 5 percent rock phosphate increased the nitrogen content of city compost by 24 to 30 percent and more favourable C: N ratios have been obtained. Available P_2O_5 content of compost was increased by 60 to 114 per cent where rock phosphate was applied and inoculated with aspergillus awamori.

Preparation of compost from enriched city garbage or otherwise is promosing, provided that financial support from government is available. However, heavy metals in sewage sludge when continuously applied in excessive quantities to farmland as organic manure could lead to problems. Monitoring for Cd, Zn, Pb. As, and Cu contents in compost is recommended.

1.2.8 Biofertilizers – A Source of Plant Nutrition

Biofertilizers have an important role to play in rainfed areas in improving the nutrient content of crop. Although *rhizobium* is the most researched and well known among the biofertilizers, there are a number of microbial inoculants with potential practical application in INM. Such inoculates could contribute to increasing crop productivity through increased biological nitrogen fixation (BNF), increased availability or uptake of nutrients through phosphate solubilization, or increased absorption, stimulation of plant growth (hormones), or by rapid decomposition of organic residues (rapid composting technology).

1.2.8.1 *Rhizobium* Inoculants

The nitrogen fixed by rhizobia benefits legume crop production in two ways: (a) by meeting most of the legume crops nitrogen needs and (b) by enriching the soil for a the benefit of subsequent crops. *Rhizobium* inoculation should be considered in al legume green manure crops to gain maximum benefit from nitrogen fixation in the shortest possible time

Azospirillum, azotobacter and *pseudomonas* inocula-tions on upland rain crops are still in their infancy and field trial results are inconclusive, although good responses to *azospirillum* and *azotobacter* inoculation of wheat, rice and sugar cane have been recorded.

Role of Rhizobium

a) Induced Nodulation

The associative Rhizobia contributes towards increase in vigour and grain yield. These are thus used in several cropping systems but the rhizoplane of the rice root system abandons the associative system to work. This flow ultimates to the induction of pseudodule/paranodule (a nodule-like stemeture in non-legume) in cereal so that the bacteria that inhibit it will provide direct benefits to the host plant harboring it. Induction could be obtained through forcefull entry of certain specific drazotrophs in presence of certain chemicals as synthetic auxins, growth promoters as cytokinin, single compounds as flavonoids that may switch on the auxin synthesis by use of cellulotytic/pectolytic enzymes along with poly etheneglycal (PEG).

b) Induction by Hydrolytic Enzymes

The first classic and landmark experiment regarding induction of paranodules in rice by rhizobia by using enzymatic removal of cell wall at root hair tips indicated that the removal of apical cell wall by mixture of cellulase and pectinase exposes the plasma-membrane and thus abolishing the specificity and can be promoted by coupling cell-wall degradation with PEG treatment as it assists bacterial entry into plant cells.

c) Induction by Synthetic Auxins

Synthetic or artificial auxins such as 2,4-D dichlorophenoxy acetic acid, paranodule formation on root of a large number of plant species including rice, wheat, maize *etc*. In presence of a rhizobia isolated from root and stem nodules of tropical *Sesbania rostrata* called *Azorhizobium caulinodans* forms paranodules in presence of 2, 4-D that show positive low rate of N fixation in crops.

1.2.8.2 Biofertilizers for Flooded Rice

Blue green algae (BGA) and azolla are most important biofertilizers for wetland rice are the water fern azolla and the blue green algae (BGA), also known as floating nitrogen fertilizer factories. Both can grow alongside paddy. Azolla can also be used for green manuring which could contribute from 20 to 60 kg per hectare N. Phosphorus is a key element and its deficiency results in poor growth and reduced N fixation (addition of 1 kg p results in fixation of 5 to 10 kg N). Azolla is considered an efficient scavenger for potassium and serves as a source of K for rice

Azolla biofertilizer technology is labour intensive. Irrigation water, phosphate fertilizer and pest control measures are necessary inputs. Nitrogen fixed by azolla or BGA becomes available to the rice crop only after its decomposition. Numerous field experiments indicated that only up to one third of the fixed N is absorbed by the following rice crop, while two- thirds remained in the soil as residual nitrogen or is lost to the atmosphere.

1.2.8.3 Phosphate Solubilizing Micro-organisms

A variety of bacteria and fungi has been identified to have the ability to solubilize and transform inorganic P from normally insoluble sources through excretion of various organic acids have been isolated. These are bacteria of the *bacillus* and *pseudomonas* app and fungi, such as aspergillus, penicillium and trichoderma spp. In addition to p-solubilization these microorganisms can also mineralize locked up organic P into soluble, plant available forms. As these reactions take place in the rhizosphere and the microorganisms bring more P into solution than they can absorb for their own growth, the surplus is available for plants to absorb. The effectiveness of these microorganisms depends on the availability of sufficient energy source. Carbon in the soil, P concentration, particle size of rock phosphate as well as temperature and moisture.

1.2.8.4 Constraints to use Biofertilizer

As a result of efforts made by various agencies during the last over one decade together with Government support, there had been considerable improvement in the popularity of biofertilizers. However, the achievement is far from satisfactory. This is simply because the crop responses to biofertilizers like organic manures are not as instant and spectacular as is possible with the use of chemical fertilizers. Also these are not available as readily as the chemical fertilizers. Thus they suffer from these and score of other constraints which are highlighted hereunder:

1.2.8.4.1 Production Constraints

Raw Material

Peat is regarded as an ideal carrier for biofertilizers. However, it is not available in India in sufficient quantities and in desirable quality. The other suitable carrier is lignite. That too is not available readily and is quite costly. In addition, materials like charcoal, press mud, agro-industrial wastages, compost *etc*. have also been tried. In tropics, agro- industrial wastes may prove better alternatives to peat *etc*. provided proliferation of contaminants is checked by a suitable anti-microbial agent, but for the cost. Storage of carrier based inoculants has also been found were prepared in a mixture of soil and charcoal or press mud and charcoal.

Specificity of Strains

Most of the biological strains of biofertilisers are soil and agro-cligical strains specific. This limits their wide spread and fool-proof use with expected performance. Therefore, region wise inadequate availability of specific strains limits their popular use.

Biological Constraints

Presence of ineffective or antagonistic strains in the bio-inoculants, which cannot be displaced easily, mars the overall efficiency of friendly bacteria in the bio-fertilizer.

Economics of Production

For the production of quality product, use of high-tech instruments and equipments is discernible. However, the low pricing structure and low off take of biofertilisers do not permit the use of sophisticated facilities in a factory. For the same reason, it is also difficult to ensure the production of truly contamination free product. Production of biofertilisers, by and large, is confined to Governmental agencies and on a small scale. The problem gets confounded with the small and extremely seasonal demand for these products.

1.2.8.4.2 Marketing Constraints

☆ Short shelf life: The shelf life of biofertilisers produced by using the common carries like peat or lignite is often less than six months. The best results are rather possible if the material is used in 3-4 months time, provided the material is not subjected to very high temperature (about 45°C) during movement and transportation to sale centers or their storage.

☆ Demand is limited because of inadequate awareness amongst the extension workers, dealers and farmers regarding cost effectiveness and usefulness of biofertilisers.

☆ Lack of interest on the part of dealers because of slow disposal rate of the product, nominal profit margin and extremely low turn over.

☆ Lack of adequate promotional and publicity support because of being a low turnover product with limited demand and profitability.

☆ Unavailability of proper transportation and storage facilities is also a major constraint for the development of their proper.

1.2.8.4.3 Precautions

☆ Keep the packet (s) of the culture at a cool and shady place, till their use.

☆ Use the material before the date of expiry mentioned on the packet.

☆ Use a biofertiliser only for the crop specified on the packet especially in the case of Rhizobium culture.

☆ Open the packet containing the culture just before use and inoculate only that much seed which could be sown immediately.

☆ Do not put culture in warm or hot water which could destroy the living bacteria contained on the biofertiliser.

☆ For seed treatment, if besides a bioinoculant, the seed is also to be treated with a fungicide or insecticide, treat the seed first with a fungicide followed by the insecticide and finally with the biofertiliser (bacterial inoculant).

☆ If the seed is also required to be treated with some mercury-based chemical, the dose of the bio-inoculant should be doubled.

☆ While using a bio-inoculant in strongly acidic or saline-alkali soils, it would be desirable to use some soil amendment along with, such as gypsum/ phosphogypsum for saline alkali soils and lime or rock phosphate *etc.* for strongly soil to maintain soil pH near neutrality (6.5 to 7.5). Else the efficiency of the biofertiliser may be adversely affected.

☆ Adequate nutrition, for instance, with phosphorus, calcium, *etc.* for the normal growth and activity of *Rhizobium* is a must and should be restored to Likewise, to meet the requirement of Ca, choice should be in favour of fertilizers like CAN or single superphosphate.

1.2.9 Legumes in Cropping System

1.2.9.1 Dual Purpose Legumes

Legumes have a long standing history of being soil fertility restorer due to their ability to obtain N from the atmosphere in symbiosis with rhizobia. Legumes can form an important component of INM when grown for fodder or grain in a cropping system. A number of leguminous crops have been evaluated for the contribution which they make in meeting the N requirements of the succeeding crop and it has been found that on an average as much as 50-60 kg N/ha may be made available. The carry over of N for succeeding cereal may be 60-120 kg in berseem, 75 kg in Indian clover, 35-60 kg in fodder cowpea, 55 kg in black gram, 60 kg in groundnut, 68 kg in gram and 50 kg in lathyrus. Grain yield of succeeding crop increase markedly when legumes precede them compared with that when cereals preceded. Pulse crops can be made more effective by inoculation with proper species of Rhizobia. In a country where the average consumption of fertilizer is very low, the residual fertility build up due to legumes is obviously a major contribution which must be fully exploited. Pulses have to be used as a source of renewable supply of N.

1.2.9.2 Legume Inter-Cropping

In situations where limitation of time may not permit, inclusion of legumes in rotations, inter-cropping of legumes may be practical solution. Sugarcane is generally planted in rows 60 to 90 cm apart. It normally take 30-45 days for germination. The initial crop growth upto 100 days after planting is slow. As a result, the lateral spread of crop foliage is not much. Secondly, green manuring prior to sugarcane is considered a wasteful practice because of population pressure on land and availability of fertilizers at cheaper rates. It is not preferred to miss a kharif crop for green manuring.

All these factors attracted the attention of scientists and farmers to discover the growing of legumes in inter-row spaces as inter crop or companion crop for grain or green manuring or fodder. Companion cropping of short duration legume crop with autumn planted can is considered a very useful innovation in economising the resources. The erect growing, short duration and dwarf varieties of legume crop are more suitable for intercropping.

Among the summer legumes, moong, urd, and cowpea and popular intercrops with spring planted sugarcane. The green manuring of intercropped legumes has been reported to increase the cane yield. The legumes, even as intercrops have shown potential for N fertilizer economy in terms of increase in the yield of sugarcane. Intercropping of legume increases N use efficiency in sugarcane by checking the losses of NO_3-N and by increasing the N uptake. The utilization efficiency of applied P is also increased by intercropping of legumes.

The success of intercropping of legumes with main crop depends upon the non-competitiveness of the intercrops and if competitive, the extent to which it may compensate the yield loss in main crop, if any, by its produce. The role of legumes has to be assessed in two ways:

i) The accretion of N

ii) Competitive loss to main crop

It may over shadow main crop at a time when the main crop is tillering and may bring initial depression in shoot population. N economy through intercropped legumes is yet to be correctly assessed in different intercropping systems. Several leguminous crops due to their small stature and short duration are ideally suited for introducing as intercrops. These intercrops may be grown for grain, fodder or as green manure. The intercrops, besides increasing the total productivity of the system, also play an important role in economising the resource use especially N. It has been estimated that with inclusion of legume in intercropping system, the extent of N addition would be of the order of 0.746 million tonnes.

1.2.9.3 Legume in Rotation

Leguminous crops are known to fix nitrogen, improve soil productivity and benefit the succeeding crop yield compared to non-legume crops in rotation, thus affecting considerable economy in plant nutrients. Studies at Pantnagar showed that wheat after inoculated soybean produced 22.4 q/ha yield compared to 13.5 q/ha after uninoculated soybean. It indicates that inoculation of legumes with suitable bacteria is desirable.

1.3 Constraints of INM

A part from several benefits and importance of INM in sustaining all over agricultural productivity and soil health nation wide, there are some constraints/limitations to use of INM practices uniquely are given as:

1.3.1 Lack of Organic Materials

Unavailability of organic materials especially animal manure and crop residues is a primary constraint in many areas.

1.3.2 Single Multiple Enterprises

Farmers who have less number of cattle may have to depend solely or mainly on fertilizers whereas farmers who practice several occupations like cash and field crops, together with dairy or livestock, poultry, fisheries enterprise *etc.*, these enterprises provide opportunities to use/recycle the wastes, manures preferentially and profitably without depending on costly purchased inputs.

1.3.3 Peri-urban/rural Differences

Developed markets encourage farmers to use fertilizers and produce more under intensive system of cropping. Even small farmers use more fertilizers inputs. However in peri-urban areas, there is also possibility for use of agro-industrial or urban municipal wastes along with fertilizers to augment soil fertility. Farmers in remote areas with poor infrastructure and without access to market and also awareness of the benefits of fertilizers may use locally available organic sources.

1.3.4 Land Tenancy

The farmers who take land on tenure basis try to harvest high yields using mineral fertilizers and irrigation to ensure a rapid returns to cover the cost of renting the land and may ignore the use of organic manures especially in cereal crop production.

1.3.5 High Cost of Organic Manures

Cost of organic manures especially the animal manures is high in peri-urban areas where these manures are preferentially used in ornamental gardens, lawns and also home gardens in raising vegetable crops.

1.3.6 Transport

Because the organic manures are bulky, it is not convenient to transport and apply them in all crops in all the seasons. So it is applied conveniently in sufficiently good amount in remunerative crops at 4-5 years interval especially in kharif crops.

1.3.7 Competitive Use of Organic Resources

A very important example of competitive use is the use of cow-dung as fuel because of the shortage of fuel wood. Similarly crop straws or stalks are used as fuel wood especially of the castor, redgram, cotton stalks. Crop residues are also very valuable animal feed. Sometimes poultry manure/droppings are used mixed with other additives and used as fish or cattle feed.

1.3.8 Pests, Diseases and Weeds

Some believe that the organic manures may carry pests, pathogens and weed seeds and propagate them in the current or following crops.

1.4 Developmental Issues

- ☆ Soil test laboratories should be strengthened and upgraded for soil and plant analysis for both macro and micronutrients.
- ☆ Greater awareness needs to be created among the farmers about the importance of soil health for monitoring soil nutrient status on regular basis through soil and plant analysis.
- ☆ For the promotion of bio-fertilizers, Government agencies, NGOs as well as Corporate sectors should be encouraged to establish bio-fertilizers production unit at least at district level. Similarly, such agencies should be provided proper support for the production of compost from city garbage.
- ☆ Ensuring availability of good quality fertilizers for micronutrient deficient areas especially zinc, sulphur *etc.*
- ☆ Fertilizer industries should also be advised to supply bio-fertilizers along with the chemical fertilizers particularly with respect to nitrogen.
- ☆ Greater awareness needs to be created among the farmers on farm resource generation and its proper recycling to serve various rural needs such as

fuel, fodder, manure, *etc.* through promotion of bio-gas plant among the farm families. Community approach may also be encouraged for the production of compost and green manure.

✰ Advantage of introduction of legumes in the cropping systems should be promoted.

✰ Special efforts to produce the seeds of efficient green manure crops are needed.

1.5 Researchable Issues

✰ "Resource management" *i.e.* strategic use of various nutrient resources is an area of challenge to soil scientists as well as agronomists and considerable part of our future research portfolio must be devoted to this issue.

✰ Optimization of various resources input combinations specific to particular soil and cropping system better utilization of input resources.

✰ Improving interactive effects of various components of INM system on various beneficial soil chemicals, physical and biological processes and nutrient use efficiency.

✰ Highly reliable diagnostic criteria of soil fertility and soil test techniques need to be developed and once, these are in use need to be refined so that their predictability could be pushed up to match the adoption of INM technologies.

✰ Need to develop model approach for computing INM for major cropping system in each of the agro-sub-ecological zones.

✰ Energy turnover on use of INM technology in different production systems.

✰ Enhancement of shelf life of bio-fertilizers, development of new strains through genetic engineering and development of easy techniques for viability test of biofertilizers.

1.6 Future Strategies

✰ Considering INM system as a mission for future agriculture, we stress that following few aspects need to be kept in mind to our vision of the future.

✰ There are many people taking about low external input sustainable agriculture (LEISA) and organic farming in the wake of sustainability debate. This, however, will not serve the future need, and INM technologies must demonstrate that a high input (inorganic and organic) and high yield agriculture can also be sustained.

✰ Since "Production sustainability" is a dynamic phenomenon, all the INM technologies aimed at sustaining productivity of our production system must have a time dimension t indicate the period up to which such technology will remain effective to maintain sustainability.

☆ INM system should not be used as complementing to synthetic fertilizers, rather it should be developed as an effective nutrient supply and management system for making best use of synthetic fertilizer and other renewable resources of nutrients.

☆ The All India foodgram production trends indicate the areas that are surplus and deficit. Movement of food grain from the surplus regions to deficit regions offsets the regional nutrient imbalances, thereby turning the intensively cultivated surplus districts regions into areas deficit in nutrient and nutrient imbalances. This trend has to be reversed through INM system.

☆ Environmental concerns should be given sufficient prominence while developing INM technologies. Ecological sustainability must be the major emphasis so that there is not adverse effect on the other life support systems on log run.

☆ A computer aided data base need to be created for all nutrient sources (mineral and organic) including their amount, composition, processing techniques, their economic value and their availability (on-farm-off-farm purchased) to help in planning strategies and managing nutrient efficiently.

1.7 Policy Issues

☆ A commission for the safer use of soil reserves and nutrient sources need to be created.

☆ Co-operative and Public Sector support for production of fertilizers to be enhanced with emphasis on improved efficiency of these sectors.

☆ There is need for uniform pricing policy on fertilizers at national level to remove the regional disparity in fertilizer use.

☆ Benefit of fertilizer subsidy to farmers must be ensured.

☆ Keeping in view the distortion of NPK ratio, there is need to develop advance planning and long-term strategy for providing adequate price support for P K fertilizer.

☆ Quality control measures must be extended to include bio-fertilizers as well for production and supply of quality bio-fertilizers.

☆ Soil test based fertilizer use is critical for which a "National Fertilizer Use Planning" is desirable.

☆ Grass root level awareness about the soil testing among farmers need to be created through panchayati system, NGOs and local agriculture supervisor. The soil as a natural resource should be considered on priority in village knowledge centers.

Chapter 2

INM Options for Some Important Crops

Research on INM in important crops and cropping system in a variety of soils has been conducted throughout the country for proper utilization of plant nutrient sources vis-à-vis sustaining crop productivity and soil health are described here serially along with suitable examples for better soil and crop management.

2.1 Rice

Rice is the most important staple food crop of the country, particularly in humid areas of Assam, Manipur, West Bengal, Orissa, Eastern U.P. and in Southern states. It may be grown in a variety of soils like clay loam, red-laterite loamy, hill mountain, black soils *etc*. The varieties of rice grown in tropics known as *Indica* and in temperate known as *Japonica*. Yield maximization experiments to identify the process and monitor the mechanisms responsible for current stagnation in yield levels clearly indicated that by resorting to conjunctive use of organo-inorganic fertilizers (Recommended fertilizer dose + FYM + $ZnSO_4$ + Biofertilizers) with optimum plant population per unit area, it is possible to maximize the grain yield of irrigated rice. It has been found that rice yield in newly cultivated areas is much higher than the traditional rice growing area, due to use of high yielding varieties, better soil and water management practices coupled with integrated nutrient management schedules. The INM options for rice crop are given in Table 2.1.

Table 2.1: Integrated Nutrient Management Options for Rice (*Oryza sativa* Linn.)

Regions/Location Soil Type/Soil Order Ecosystem/ Agro-climatic Regions	INM Options
Rajendranagar (A.P.) Alfisols Semi-arid	Substitution of recommended nitrogen either 50 or 25 percent by use of green manure *Glyricidia* gave an additional yield of rice @ 2.71 q/ha and was at par with 100 percent recommended dose of nitrogen fertilizer.
Jabalpur (M.P.) Clay soil Sub-humid	Incorporation of 5 t/ha of FYM and 40 kg N in three splits of 15+15+10 kg and 15 kg P_2O_5/ha help in increasing yield and sustaining soil health.
Pusa, Samastipur (Bihar) Silty clay loam Sub-humid	Application of 75: 45: 30 (N: P: K kg/ha)+ *prickly sesban* @ 14 t/ha proved effective in increasing rice yield.
Klyani (W.Bengal) Alluvial Sub-humid	Use of nitrogen 60 kg, phosphorus 30 kg and potassium 30 kg per hectare along with 10 t FYM/ha increased yield and soil health.
Imphal (Manipur) (Submerged)	Use of *Azotobactor* at root dip, tillering and booting stage with nitrogen application by urea found beneficial in submerged conditions.
Madurai (Tamil Nadu) Semi-arid	*In situ* incorporation of green manure, Daincha @ 12 t/ha along with 50 percent recommended nitrogen with inoculation of *Azospirillum* increased rice yield.
Chandrashekarpur (Orrisa) Rainfed lowland	Application of 75 percent (60: 30: 30, N: P: K kg/ha) of recommended dose based on fertilizer along with green manure dhaincha increased rainfed lowland rice yield.
Annamalinagar (Tamil Nadu) Low land rice Semi-arid	Lowland rice yield increased by application of 180: 57: 57 (N: P: K kg/ha) along with pressmud @ 25 t/ha. Seed treatment with KHz Po4 = 100 percent N: P: K+use of *Azospirillum* and *phosphobacteria* gave highest rice yield (5.72 t/ha) at Annamalinagar.
Larnoo (J&K) Alfisol Semi-arid	Incorporation of 20 t FYM/ha along with N: P: K (80: 60: 40) kg/ha proved beneficial in sustaining crop yield and soil health.
Allahabad (U.P.)	Use of flyash @ 40 t/ha along with recommended dose of fertilizer and inoculation with *Azospirillum* proved effective.
Rahchi (Jharkhand) Oxisols	Application of 40: 30: 20 NPK kg/ha in upland rice proved effective3 in increasing yield as well as nutrient and water use efficiencies.
Varanasi (U.P.) Entisols	Application of 40: 30: 20 kg NPK/ha increased yield of upland rice.

Regions/Location Soil Type/Soil Order Ecosystem/ Agro-climatic Regions	INM Options
Coimbatore (Tamil Nadu) Sandy clay loam Semi-arid	Incorporation of green manure (GM) *Sesbania aculeata* @ 6.25 t/ha or FYM @ 12.5 t/ha along with *Azospirillum* 2 kg/ha and nitrogen @ 150 kg/ha with zinc sulphate @ 25 kg/ha help in sustaining crop yield as well as soil health.
Bangalore (Karnataka) Sandy loam Semi-arid	Use of STCR dose (75 per cent inorganic + 25 per cent organic (4 t FYM/ha) helps in increase of productivity of rice.
Nellore (Andhra Pradesh) Semi-arid	Incorporation of green manures Sunnhemp (5 t/ha) plus soil test based dose of fertilizer increased rice yield.
Thandamuthur Udumalpet (Tamil Nadu) sandy loam Arid	Green manure such as *Calotropis* and Kolingi with inorganic fertilizers improved productivity of rice and soil.
Bhubaneswar (Orissa) Sandy loam Sub-humid	Incorporation of 5 t FYM/ha 127: 0: 41.4 N: P_2O_5: K_2O/ha increased yield of rice and improved soil properties.
Durg (Chhattisgarh) Vertisols Sub-humid	Combined application of mineral fertilizers + 5 t/ha FYM increased seed yield and soil health.
Ropar (Punjab) Sandy loam Arid	Application of 124 kg N/ha + 13 t/ha FYM gave highest grain yield.
Shalimar (J&K) Semi-arid Silty clay loam	The application of spent mushroom compost (SMC) enhanced the yield of paddy by 19 per cent over control and was found superior to FYM.

2.2 Wheat

Among the cereals, wheat is the most widely cultivated crop in the World and in India. At present India produces more than 72 million tonnes of wheat, which is 11 times higher relative to the same during 1950-51. At present wheat production in India is more than that of United States. The contribution of our country to global wheat production has increased from 10 percent in nineties to 12 percent at present. In India, it is grown in zones like North Eastern Plains, North Western Plains, Central Zone, Peninsular Zone, Northern hilly areas. It was observed that 9.2 per cent of increased grain yield was due to use of biofertilizers with average extra yield of 330 kg per ha. As so far integrated nutrient management options are concern in wheat crop, it is recorded at various places that the green manuring, use of caster cake, use of *Azotobactor*, incorporation of FYM, vermicomposting and spent mushroom compost increased wheat grain yield and post harvest available plant nutrients. There frequent uses improving physical and biological properties of wheat growing soils. The exmples of INM options in wheat crop are given in Table 2.2.

Table 2.2: Integrated Nutrient Management Options for Wheat (*Triticum aestivum* Linn.)

Regions/Location Soil Type/Soil Order Ecosystem/Agro-climatic Regions	INM Options
Hoshiarpur (Punjab) Entisols Arid	Growing of green manure crop of sunhemp in light textured soils, which are generally kept fallow during kharif is an important practice. Incorporation of GM crop in the field in the middle of August and raise wheat with the application of 80 kg nitrogen/ha in two equal splits.
SK Nagar (Gujarat) Loamy sand Arid	Application of 90 kg N/ha through fertilizers + 30 kg N/ha through castor cake gave highest grain yield
Hisar (Haryana) Sandy loam Arid	Recommended dose of 120 kg N + 60 kg P_2O_5 + 60 kg K_2O/ha + inoculation of wheat seeds with *Azotobacter* sustained crop productivity.
Gurgaon (Haryana) Loamy sand Semi-arid	Incorporations of FYM @ 10 t/ha + N120 +P26 +K50 increased grain yield.
New Delhi Sandy loam Semi-arid	Incorporation of 90 kg N/ha + FYM @ 10 t/ha increased grain yield.
Ahmednagar (Maharashtra) Sandy loam Semi-arid	20 per cent reduced dose of fertilizer plus FYM @ 2 t/ha and *Azotobactor* 20 per cent reduced dose + 5 t FYM/ha + *Azotobacter*.
Hisar (Haryana) Sandy loam Arid	Use of vermicompost @ 15 t/ha + recommended dose and NPK increased yield of wheat.
Hisar (Haryana) Sandy loam Arid	Application of 100 per cent NPK (150: 60: 60) kg/ha + 0.75 per cent spent mushroom compost (SMC) improved yield and soil health.
Udaipur (Rajasthan) Sandy loam Semi-arid	Cross sowing of wheat + 90 kg N + 40 kg P_2O_5 + 25 kg $ZnSO_4$/ha increased grain yield of wheat and improved nutrient uptake.

2.3 Sorghum (Jowar)

It is known for its drought tolerance and is most popular food/fodder crop in dryland areas of Central and North Indian Zones. Sorghum is grown about 99.8 lakh ha area. The major jowar growing states are M.P., A.P., Karnataka, Maharashtra,

Gujarat, Rajasthan. Although jowar is a kharif season crop however, in some central and southern part of the country, it is grown as a major rabi crop. It can be grown on a variety of soil types but the clayey loam soil rich in humus is found to be the most ideal soil. It may tolerate mild acidity to mild salinity under pH 5.5 to 8.0. The INM options in sorghum are given in Table 2.3 revealed that the balanced and conjoint use of fertilizers, organics and biofertilizers gave highest fodder and sorghum grain yield at Udaipur, Hyderabad, Paiyur, Faizabad and Akola.

Table 2.3: Integrated Nutrient Management Options for Sorghum (*Sorghum vulgare* Pers.)

Regions/Location Soil Type/Soil Order Ecosystem/ Agro-climatic Regions	INM Options
Udaipur (Rajasthan) Sandy loam Semi-arid	Incorporation of 5 t FYM/ha + N (80 per cent) + 75 per cent (Fertilizer) + *Azospirillum* + PSB sustained crop productivity and soil health.
Udaipur (Rajasthan) Sandy loam Semi-arid	Incorporation of FYM @ 10 t/ha + 75 per cent RDF + *Azospirillum* + PSB gave highest grain yield of sorghum.
Hyderabad (A.P.) Alfisol Semi-arid	(i) Application of 2 t/ha *glyricidia* loppings along with 20 kg nitrogen through urea. (ii) Application of 4 t/ha FYM + 2 t/ha *glyricidia* loppings or (iii) Application of 4 t/ha FYM + 20 kg N/ha through urea. Apart from supplying nutrients, the practices will help in improving soil health by increasing soil organic matter.
Paiyur (Tamil Nadu) loamy sand Semi-arid	Recommended mineral fertilizers + FYM @ 10 t/ha or 75 per cent recommended inorganic fertilizers + biofertilizers increased sorgham grain yield and sustained soil health.
Faizabad (U.P) Inceptisol Semi-arid	Application of 15 kg N through compost along with 20 kg nitrogen through inorganic fertilizer increased yield.
Akola (Maharashtra) Vertisols Semi-arid	Incorporation of 9.45 t/ha of *leucaena loppings* during fallow provides 25 kg N/ha to rati sorghum.
Akola (Maharashtra) Vertisols Semi-arid	Application of 50: 50: 50 kg NPK/ha through pressmud cake and 50 kg N through urea as top dressing increased sorghum yield.

2.4 Maize

Maize is an important staple food crop in India which is grown on plains, hills or mountainous areas having moderate rainfall. In India the major area of the crop is confined to Gujarat, Rajasthan, Punjab, Haryana, M.P., U.P., A.P., H.P., Jammu and Kashmir and Bihar. Deep, fertile, rich in organic matter and well drained soils are most preferred for the maize, however may be grown on a variety of soil types. The soil should be medium textured with good water holding capacity. The exploitation of soil as a natural resource has been tremendously increased in the modern agriculture. It is an urgent need to restore the inherent capacity of a soil to supply plant nutrients besides high yield of maize. Integrated nutrient management options for maize are given in Table 2.4 suggested that the use of organics may substituted 25 to 50 percent of the total nutrient needs of maize crop at Madurai, Coimbatore, Palampur and Meerut.

Table 2.4: Integrated Nutrient Management Options for Maize (*Zea mays* Linn.)

Regions/Location Soil Type/Soil Order Ecosystem/ Agro-climatic Regions	INM Options
Satpura (M.P) Lighter soil of satpura plateau	Combine use of 5 t/ha vermicompost + NPK (RDF) recorded highest grain yield (24.35 q/ha)
Chindwara (M.P) Sub-humid	Application of 90: 40: 30, N: P_2O_5: K_2O kg/ha + FYM 10 t/ha increased grain yield and sustained crop productivity.
Madurai (Tamil Nadu) Sandy clay loam Semi-arid	Application of 75 per cent RDF through fertilizer + 5 t FYM/ha increased higher grain yield of maize.
Larnoo (J&K) Alfisol	Use of 80 kg N + FYM @ 10 t/ha recorded highest grain yield.
Rakh Dhiansar Entisol	Application of 60-40-20 kg NPK/ha + 20 kg/ha zinc sulphate improved yield in rainfed conditions.
Coimbatore (Tamil Nadu) Medium black	Application of 75 per cent N through vermicompost along with 25 per cent N through neem cake improved yield and soil health.
Palampur (H.P.) Clay loam	Application of FYM can substitute 50 per cent of recommended nitrogen helps in economizing 50 per cent of recommended NPK in maize.
Modipuram (U.P) Sandy loam Semi-arid	Application of 50 per cent recommended dose through fertilizer and remaining 50 per cent N as FYM helps in sustaining crop productivity and soil health.

2.5 Pearlmillet (Bajra)

Bajra is grown for grain as well as forage purposes. It is a popular crop grown in areas with less moisture conditions. It is a seasonal kharif crop but in Central and Southern India and Indogangetic areas, the crop is grown during summer. It is grown on variety of soils such as sandy loams of U.P. and Punjab and light soils of Rajasthan and North Gujarat, heavy clays of A.P., Tamil Nadu and very light soils of Maharashtra and shallow black, red and light soils of Deccan and Southern India. Green manuring along with recommended fertilizer doses found beneficial for improving physico-chemical properties of soil besides yield increment. Some INM options for pearlmillet crop are given in Table 2.5.

Table 2.5: Integrated Nutrient Management Options for Pearlmillet (*Pennisetum typhodeum* Linn.)

Regions/Location Soil Type/Soil Order Ecosystem/ Agro-climatic Regions	INM Options
Hisar (Haryana) Sandy loam Arid	FYM @ 2.5 tonnes/ha + 20 kg N/ha + seed treatment with biofertilizers help in increasing yield and sustained soil health.
Durgapura (Rajasthan) Sandy loam Semi-arid	Application of 5 t/ha vermicompost or 100 per cent RDF proved beneficial in rainfed pearl millet crop.
Jamnagar (Gujarat) Medium black soils	Incorporation of 5 t FYM/ha + 40 kg N/ha increased grain yield of bajra
Hisar (Haryana) Sandy loam	Application of 120 kg N + 60 kg P_2O_5-along with vermicompost @ 5 t/ha recorded highest seed yield (2270 kg/ha) of rainfed pearl millet.
Rahuri (Maharashtra) Medium black Semi-arid	Incorporation of green leaf manuring one month before sowing (5000 kg/ha subabul or *glyricidia)* or intercropping with cowpea and sunnhemp improved soil physico-chemical properties of soil.
Agra (U.P.) Entisol and Inceptisol	Application of 40 kg N/ha through fertilizer help in increasing yield.
Hisar (Haryana) Sandy loam Semi-arid	Application of 94: 60: 0: 25 kg N: P_2O_5: K_2O: $ZnSO_4$ along with incorporation of 1.5 t FYM/ha in kharif pearl millet recorded highest grain yield.
Coimbatore (Maharashtra) Sandy loam Semi-arid	Use of 75 kg N/ha as urea + FYM @ 2.5 t/ha increased grain yield.
Coimbatore (Tamil Nadu) Clay loam Semi-arid	Use of *Azospirillum* + 100 per cent NP + *phosphor-bacterium* contained crop and soil productivity.
Leh (J&K) Sandy loam Sub-humid	Application of 60 kg N/ha along with FYM 10 t/ha improved grain yield and soil health.

2.6 Soybean

Soybean is grown on a variety of soil types but fertile well drained soils are better. It is grown as kharif crop besides it grows partly in spring season. It prefers neutral soil having a pH between 6.5 and 7.5. It can tolerate mild salinity but acidity is deadly harmful to the crop. It can be grown on light to black cotton soils of M.P. Seed treatment of soybean with *Rhizobium* and PSB found beneficial in increasing yield at Rewa in vertisols (Table 2.6). The integrated use of nutrient surces may save upto 25 percent of fertilizers.

Table 2.6: Integrated Nutrient Management Options for Soybean (*Glycine max* Merr.)

Regions/Location Soil Type/Soil Order Ecosystem/ Agro-climatic Regions	INM Options
Coimbatore (Tamil Nadu) Sandy loam Semi-arid	Application of 75 per cent recommended nitrogen along with biofertilizers found beneficial for better yield and soil health.
Sagar (M.P) Medium black Semi-arid	Application of 100 per cent N: P: K (20: 80: 20kg/ha) recorded highest seed yield/plant.
Indore (M.P) Vertisol Semi-arid	Incorporation of FYM @ 6 t/ha along with 20 kg nitrogen and 30 kg phosphorus/ha increased seed yield and soil health in indore region.
Rewa (M.P) Vertisol Semi-arid	Seed treatment of soybean with *Rhizobium* bacteria and phosphate solubilizing bacteria (PSB) helps in increasing yield of soybean.
Bhopal (M.P) Black soil Sub-humid	Balanced fertilization based on soil test along with incorporation of FYM @ 4 t/ha increased yield of soybean and sustained soil health.
Udhamsingh Nagar (Uttaranchal) Clay loam Sub-humid	Application of 25 kg N + 5 tonnes FYM/ha or 25 kg N + 1 tonnes neem cake/ha gave highest yield.
Bapatla (A.P.) Sandy loam Sub-humid	Use of biogas slurry 15 t/ha + 50 kg N/ha as urea help in sustaining crop productivity.
Udaipur (Rajasthan) Sandy loam Semi-arid	Seed treatment of soybean with *Rhizobium* culture @ 5 g/kg of seed arid soil inoculation of phosphate solubilizing bacteria (PSB) @ 500 g/ha by mixing with 50 kg of well decomposed FYM at the time of sowing increased fertilizer uptake and yield of soybean and sustained soil health.

2.7 Groundnut

It is the most important oilseed crop grown in India occupying an area of about 7.572 lakh ha. It is grown mainly in Gujarat, Karnataka, Andhra Pradesh and Tamil Nadu states. The crop requires sandy to loamy soils having drainage facilities and uniform topography. Groundnut also has the commercial value for supplying the raw materials for industrial uses. Integrated nutrient management practices for groundnut crop observed at Rajkot, Mizorum, Secundrabad, Kanpur, Uridhachalam, Sangli, Anantpur and Praphani are given in Table 2.7 for yield and soil health improvement.

Table 2.7: Integrated Nutrient Management Options for Groundnut (*Arachis hypogea* Linn.)

Regions/Location Soil Type/Soil Order Ecosystem/ Agro-climatic Regions	INM Options
Rajkot (Gujarat) Vertisols Arid	Application of NPK (6: 12: 0 kg/ha) along with mulching of sunhemp in between rows plus *Rhizobium* and PSB treatment gave higher pod yield.
Mizorum (Meghalaya) Hapludult Sub-humid	Application of 100 per cent NPK (Full recommended dose) combined with pig manure @ 5 t/ha recorded 20.02 q/ha yield and improved physical properties of groundnut growing soils.
Secunderabad (Telangana) Sandy loam Semi-arid	Application of 100 per cent NPK (Full recommended dose) combined with inoculation *Rhizobium* @ 2 kg/ha and *phosphobacteria* @ 2 kg/ha as seed treatment recorded higher yield and sustained soil health.
Kanpur (U.P.) Sandy loam Semi-arid	NPK (20: 30: 45 kg/ha) with incorporation of FYM @ 10 t/ha + gypsum application @ 300 kg/ha for groundnut crop proved beneficial.
Vridhachalam (Maharashtra) Alfisols Semi-arid	Incorporation of FYM @ 10 t/ha combined with biofertilizer (*Rhizobium* + *phospho-bacterium*) helps in increasing crop productivity and soil health.
Sangli (Maharashtra) Black soils Semi-arid	Incorporation of FYM @ 2 t/ha + 75 per cent recommended dose of NPK to kharif groundnut increased pod yield.
Anantpur (Maharashtra) Alfisol	Use of *Rhizobium* @ 50 gram/ha with N, P_2O_5, K_2O (20: 40: 40) kg/ha increased groundnut yield at Anantpur
Parbhani (Maharashtra) Clay Semi-arid	Application of 50 per cent NPK through chemical fertilizers and 50 per cent NPK through inorganic plant nutrient sources (FYM + *Azotobactor* + cow-dung urine slurry + phosphate solubilizing bacteria) sustained soil and groundnut production.

2.8 Sunflower

Sunflower is an oilseed crop and largely cultivated throughout India. It needs warm humid climate during vegetative growth, bright sunny days during flowering and maturity but, prefers light textured well drained soils. The nutrient requirements of oilseeds like sunflower in general, is high for all the nutrients which need to be supplied in adequate and balanced quantities for high yield. Use of biofertilizer like *Azotobacter, Azospirillum* and phosphate solubilizing microbes can also play an important role in achieving economy in chemical fertilizer use. Integrated nutrient management options for sunflower crop are given in Table 2.8 advocated that the use of biofertilizer and FYM can substitute 25 percent of the total nitrogen requirement of sunflower.

Table 2.8: Integrated Nutrient Management Options for Sunflower (*Halianthus annus* Linn.)

Regions/Location Soil Type/Soil Order Ecosystem/ Agro-climatic Regions	INM Options
Hisar (Haryana) Sandy loam Semi-arid	Application of 30 kg N/ha through fertilizer along with incorporation of vermicompost @ 7.5 t/ha with *Azotobactor* increased crop productivity and soil health.
Dharwad (Karnataka)	Use of 100 per cent recommended dose of fertilizer along with either FYM or vermicompost or poultry manure increased sunflower seed yield.
Hisar (Haryana) Sandy loam Semi-arid	Application of 30 kg N/ha with incorporation of FYM 10 t/ha increased sunflower yield.
Coimbatore (T.N.) Sandy loam Semi-arid	Application of 75 per cent recommended nitrogen through fertilizer along with use of biofertilizers *(Azotobactor* and PSB) helps in increasing yield and soil health.
Hyderabad (A.P.) Sandy loam Semi-arid	Use of P-enriched FYM (phosphocompost) + *Azospirillum* + *phosphobacteria* with 0.2 per cent Borax helps in increasing yield of sunflower
Shalimar (J&K) Silty clay loam	Use of 75 per cent recommended dose of NPK along with 10 t/ha FYM proved beneficial.
Hyderabad (A.P.) Alfisols	Application of 5-10 t/ha FYM in conjunction with 50 per cent recommended dose of NPK offer enhanced yield and nutrient use efficiency.

2.9 Cotton

It is one of the most important fibre yielding crop in India. The cotton contribution to our country's economy is of high significance. It is grown in three-agro-ecological zones *viz.*, Northern (Punjab, Haryana and Rajasthan), Central (Gujarat, Maharashtra and M.P) and Southern (A.P., Tamil Nadu and Karnataka).

All having the different species compositions and their zone specific problems. Integrated nutrient management aimed at sustainable crop production by combined use of inorganic fertilizers, organic manure and biofertilizers to promote the inherent nutrient supplying capacity of the soil are need of the time. A few INM options for cotton are given in Table 2.9 for increasing cotton productivity and sustaining soil health.

Table 2.9: Integrated Nutrient Management Options for cotton (*Gossypium* spp.)

Regions/Location Soil Type/Soil Order Ecosystem/ Agro-climatic Regions	INM Options
Bhawaniputna (Orrisa) Clay soil Sub-humid	Application of N,P and K @ 100, 50 and 50 kg/ha in conjunction with *Azotobacter* and PSB @ 5 kg/ha each along with incorporation of FYM @ 10 t/ha increased cotton boll yield and sustained soil health.
Hisar (Haryana) Sandy loam Semi-arid	Application of 50 per cent NPK recommeded dose through fertilizers and 50 per cent nitrogen through FYM recorded highest gross return in cotton.
Akola (Maharashtra) Clay loam Semi-arid	Application of 50 kg N and 25 kg P_2O_5/ha through urea and single superphosphate (SSP) along with FYM @ 10 t/ha helps in increasing yield and soil health.
Nagpur (Maharashtra) Medium black soil Semi-arid	Use of biofertilizer strains of *Azotobactor*, PSB and VAM based INM packages proved beneficial to crop and soil in cotton black soils.
Nagpur (Maharashtra) Black soils Semi-arid	Application of 50 per cent NPK of recommended dose alongwith incorporation of FYM @ 10 t/ha helps in sustaining crop and soil productivity.
Parbani (Maharashtra) Vertisols Semi-arid	Combined application of 50 per cent recommended dose through chemical fertilizers and 50 per cent recommended dose through FYM and seed treatment with *Azotobactor* increased cotton boll yield and sustained soil health.

2.10 Mustard

India is the largest producing country of mustard in the world occupying an area of about 6.5 lakh ha. The major mustard producing states are Rajasthan, U.P., Punjab, Haryana, Assam, Bihar, M.P., West Bengal and Orissa. It is a dry season crop which requires cool and dry weather with bright sunshine. It is grown under rainfed as well as irrigated conditions. Research on INM indicates that the 25 to 50 percent of the NPKS needs can be substituted by combination of organic, inorganic and biofertilizers. Besides INM, it is necessary to apply 20 - 40 kg/ha sulphur and 25 kg zinc sulphate in the sulphur and zinc deficient soils. A few examples of INM options are given in Table 2.10.

Table 2.10: Integrated Nutrient Management Options for Mustard (*Brassica* sps.)

Regions/Location Soil Type/Soil Order Ecosystem/ Agro-climatic Regions	INM Options
Udaipur (Rajasthan) Sandy loam Semi-arid	Application of 75 per cent NPKS recommended dose based on soil test along with incorporation of FYM @ 10 t/ha or 75 per cent NPKS based on soil test alongwith use of biofertilizer *(Azotobactor* +PSB) helps in increasing seed and oil yield of mustard and sustained sole health. Both INM practices proved effective in increasing soil physico-chemical and microbial properties of soil.
Kalyani (W.Bengal) Alluvial Sub-humid	Combined application of N: P: K (20: 100: 100 kg/ha) and FYM @ 10 t/ha increased seed yield of mustard and improved soil physical properties.
Jamnagar (Gujarat) Medium black	Application of 50 per cent NPK along with seed treatment with biofertilizers (PSM + *Azotobactor)* gave highest yield of mustard.
S.K.Nagar (Gujarat) Calcareous and saline soil arid	On-farm trials conducted on saline farmers field showed that application of caster cake 1 t/ha increased mustard yield by 14.1 per cent over farmer practice.

2.11 Sugarcane

This sugarcane is the main source of sugar in India. It is commercially grown in Maharashtra, Karnataka, Tamil Nadu, Punjab, Haryana, U.P., Bihar, Assam, West Bengal, Orrisa and A.P. It requires warm and humid climate for its growth, however, the crop needs high temperature, bright sunlight and high humidity for growth but cool, dry bright sunshine and frostless night during maturity. The crop needs heavy soil, rich in organic matter and plant nutrients, but should be free from salt ideally green manuring and use of sugarcane trash in sugarcane crop helps in increasing urea utilization when combinely use with fertilizers. INM options are given in the Table 2.11 for further reading.

Table 2.11: Integrated Nutrient Management Options for Sugarcane (*Saccharum officinarum* Linn.)

Regions/Location Soil Type/Soil Order Ecosystem/ Agro-climatic Regions	INM Options
Gangavati (Karnataka) Vertisols Arid	Application of 125 per cent recommended dose of fertilizer along with pressmud @ 3.5 t/ha increased cane yield.
Madurai (Tamil Nadu)	Incorporation of composted trash @ 5 t/ha + *Azotobactor* + recommended levels of NPK improved cane yield and soil physical properties of sugarcane growing soils.

Regions/Location *Soil Type/Soil Order* *Ecosystem/* *Agro-climatic Regions*	*INM Options*
Shahjahanpur (U.P.)	Green manuring is an important practices in these areas, and 50 per cent of the benefit of green manuring could be obtained in the form of increased sugarcane yield.
Modipuram (U.P.) Sandy loam Semi-arid	When organic manures like FYM, crop residues, poultry manure *etc.* are used along with fertilizers, the efficiency of urea utilization by sugarcane crop increase and help in better soil health

Chapter 3

Pulses

Pulses are important for protein in our daily diet. The protein from pulses is easily digestible and relatively cheaper and has higher biological value besides, pulse crop possesses ability to support soil bacteria (*Rhizobium*) in their root nodules which fixes atmospheric nitrogen and enrich soil fertility. Most pulse growing states are Karnataka, A.P., Maharashtra, Haryana, Punjab, M.P, U.P., Uttranchal, Chattisgarh and Rajasthan. Black gram is grown as a spring and summer crop. Green gram is grown as rainfed, during kharif and summer. It grows well on alluvium, red and black cotton soils. Bengal gram is grown in irrigated rabi besides in dryland areas. It is also grown as mixed with cereal or millet but under assured irrigation it is grown as a pure crop. Besides various integrated nutrient management options as suggested in Table 3.1 at various places, it is necessary to use micronutrients like Zn, B, Mo and Fe helps in improving productivity in the following manner.

☆ Foliar spraying of 0.5 kg $ZnSO_4$/ha with 0.25 lime is effective in correcting Zn deficiency.

☆ Mo deficiency can be corrected by applying 1 kg sodium molybdate/ha.

☆ Soil application of $ZnSO_4$ @ 25 kg/ha to one crop on Zn deficient soils is helpful to both the crops of a pulse based cropping system.

☆ Foliar spray of B @ 1.0 – 1.5 kg B/ha or soil application of 4 kg borax/ha enhances pod yield on B deficient soils.

☆ Spray of 2 per cent $FeSO_4$ to recoup from Fe deficiency.

Table 3.1: Integrated Nutrient Management Options for Pulse Crops

Pulses	INM Options
Chickpea *(Cicer arietinum* Linn.)	(i) At Junagadh (Gujarat) application of 20 kg N + 40 kg P_2O_5/ha + *Rhizobium* inoculation helps in increasing pod yield and sustaining soil health in medium black soils.
	(ii) At Sikundrabad (A.P) in Semi-arid ecosystem, application of 100 per cent NPK + *Rhizobium* 2 kg/ha + phosphobacteria @ 2 kg/ha improved the yield and sustained soil health in sandy loam soils.
	(iii) At Udaipur (Rajasthan) in sandy loam soils, application of N(50 per cent) + P_2O_5 (50 per cent) through inorganic fertilizer along with *Rhizobium* + PSB + VAM sustained crop productivity and soil health.
	(iv) At Bellary in Vertisols, application of 15 kg N through green leaf + 10 kg N through inorganic fertilizer increased pod yield of chickpea.
Pigeon pea *(Cajanas cajan* Milsp)	(i) At Akola (Maharashtra) in vertisols, application of FYM @ 5 t/ha + 40 kg P2O5/ha + microbial culture @ 15 kg/ha to pigeon pea.
	(ii) At Sehore (M.P.) use of 50 per cent N through urea and 50 per cent N through pressmud/cotton cake/castor cake compost + recommended P_2O_5 and K_2O increased pod yield of pigeon pea in Vertisols.
Greengram *(Vigna radiata* Boxb.)	Application of 75 per cent recommended dose through inorganic fertilizer combina-tion with use of poultry manure (25 per cent) dose gave the highest yield (11.36 g/ha) at Allahbad (U.P.) in sandy loam soils.
Peas *(Pisum sativum* Linn.)	At Palampur (H.P) in clay loam soil of Sub-humid ecosystem, application N: P: K (20: 60: 30 kg/ha) along with incorporation of FYM @ 20 t/ha improved pea yield and soil properties.
Cow pea *(Vigna sinensis* Savi)	Application of 75 per cent recommended N and biofertilizer increased cowpea yield at Coimbatore (Taminadu) in sandy soil.
Blackgram *(Vigna mungo var. radiatus* Linn.)	(i) At Coimbatore (Tamil Nadu) application of 75 per cent P_2O_5 as *Tunisia* rock phosphate + vermicompost @ 2 kg/ha + phosphobacteria @ 2 kg/ha improved crop and soil productivity.
	(ii) At Marutera (A.P.) in vertisols inoculation of VAM + 50 kg P_2O_5/ha
	(iii) increased pod yield of blackgram.
Greengram *(Vigna radiata* Roxb.)	(i) Application of 10 kg N/ha through urea under conventional tillage along with incorporation of 2 t/ha FYM increased mungbean yield at Hyderabad (A.P) in Alfisols.
	(ii) Incorporation of FYM @ 2 t/ha along with *Glyricidia loppings* @ 1 t/ha under reduce tillage improved crop productivity and soil health.

Chapter 4

Vegetables

Vegetables are commercially grown in almost all parts of the country and linked by both rich and poor. Presently vegetable cultivation occupies 6.09 million hectares area with an annual production of 84.8 million tonnes. Potato, tomato, brinjal, cauliflower and cabbage are most importantly prefered by public besides okra, onions and pea. The leading states are U.P., Bihar, Orrisa in north India and Maharashtra, Tamil Nadu and Karnataka in South India. Farmers tend to over use of fertilizers in vegetables. This practice damages their quality. According to the nutrient supply of each crop is to be balanced by the use of organic manures and fertilizers. For sustainable vegetable production, there must be an optimum nutrient balance. Continuous use of high nitrogenous fertilizers alone cause sharp reduction in soil organic matter. Decline in soil organic matter may be reestablished by the balance use of NPKS in conjunction with annual application of 10-15 t/ha FYM. INM options in various vegetables are given in Table 4.1 for increasing yield and maintaining soil health. The general INM recommendations in vegetable are follows as given below:

☆ Addition of organic manure as well as decomposed cowdung, poultry litter and chemical fertilizer recommended.

☆ Liming at 2 ton/ha every 1 to 2 years is recommended. Lime should be added 2-3 weeks before the addition of chemical fertilizer in acidic soils.'

☆ Crop rotation should be practiced.

☆ All N fertilizers covered with the layers of soil to prevent volatilization losses.

Table 4.1: Integrated Nutrient Management Options in some Vegetable Crops

Vegetables	INM Options
Brinjal *(Solanum melongena)*	Application of 50 kg nitrogen through urea along with 50 kg nitrogen through poultry manure per hectare help in increasing yield of brinjals and sustained soil health.
Okra *(Abalmosms caculantus)*	(i) Application of 20 kg nitrogen by Ammonium sulphate with 20 kg nitrogen through poultry manure per hectare increased fruit yield of okra. (ii) At Knonkan (Gujarat) incorporation of FYM @ 15 t/ha along with N, P_2O_5 and K_2O (100: 50: 20) or biofertilizer (Plantrich) @ 375 kg/ha increased fruit yield and help in sustaining soil health. (iii) Combination of organic FYM @ 10 t/ha) and inorganic (50: 40: 25 NPK kg/ha) found beneficial in increasing yield.
Tomato *(Lycopersicon esculentum* Mill)	Incorporation of FYM @ 40 t/ha along with half dose of NPK (75: 30: 30) kg/ha substituted 50 per cent of recommended dose of fertilizer.
Onion *(Allium cepa* L.)	At Wadura of Jammu and Kashmir application of 52.5 kg nitrogen per hectare with use of *Azotobactor* biofertilizer recorded higher onion yield.
Garlic *(Allium sativum* L.)	At Shalimar Srinagar in silty clay loam soil inoculation of seed with *Azotobactor* and phosphobacteria + application of 75 kg nitrogen and 45 kg phosphorus helps in better yield and soil health.
Chilli *(Capsicum annum* L.)	At Dapoli (Maharashtra) post or FYM or Glyricidia @ 10 t/ha + 50 per cent Recommended dose of fertilizer proved beneficial.
Cabbage *(Brassica olerecia* var capitata)	(i) Incorporation of FYM @ 10 t/ha along with fertilizers (N: P_2O_5: K_2O) 150: 60: 60 kg/ha increased cabbage yield and sustained soil health at Shalimar in silty clay loam soil. (ii) At Coimbatore (Tamil Nadu) in sandy clay soil application of 75 per cent recommended nitrogen along with inoculation with biofertilizer proved beneficial.
Potato *(Solanum tuberosum)*	(i) Application of 75 per cent recommended fertilizer along with incorporation of 20 t/ha FYM increased tuber yield of potato in clay soils of Palampur. (ii) At Kalyani (West Bengal) in alluvial soil application of N: P: K (60: 30: 30) kg/ha along with incorporation of FYM @ 10 t/ha recorded higher tuber yield. (iii) At Shimla (H.P.) incorporation of vermicompost @ 5 t/ha found beneficial and recorded (65 q/ha tuber yield) proved the importance of vermicompost in tuber crops. (iv) At Shalimar in silty clay loam soil application of 75 per cent recommended dose of NPK along with treatment of tuber with 1 per cent urea and 1 per cent sodium bi-carbonate help in better quality production of potato. (v) At Shalimar in silty clay loam soil incorporation of FYM @ 30 t/ha + fertilizer N: P: K @ 120: 60: 60 kg/ha increased tuber yield.
Broccoli *(Brassica oleracea* L. var. Halicaplencic)	Application of FYM + Digested sludge (each @ 10 t/ha) and seedling inoculation with VAM only noticed significant improvement in fresh and dry weight of head, head yield at Varanasi.

Chapter 5

Spices

India is the major producer and exporter of pepper, ginger, turmeric and seed spices. Integrated nutrient management is one of the important factor besides good seed, water, pesticides, which contributes in higher yield of cumin crops, coriander, ginger, fennel and turmeric crop. Use of organic manures in spice cultivation is must for maintaining quality as well quantity of spices. Biofertilizer uses has also increased in spices in recent years and fruitful results were observed in several states. A few example of INM are given in Table 5.1 advised to use all plant nutrient sources incombination gave better yield of spices. Use of organic manure, mulching, crop residues, green manuring and composting are the general INM recommendations in vegetable crops for higher yield and good soil health.

Table 5.1: Integrated Nutrient Management Options for some Spices

Spices	INM Options
Cumin crops (*Cuminum cyminum* L.)	At Ajmer (Rajasthan) use of *Azospirillum* + Sheep manure @ 10 t/ha + vermicompost @ 7.5 t/ha increased cumin yield and improved soil physical and microbiological properties.
Coriander (*Coriandrum sativum*)	At Kota (Rajasthan) application of 50 per cent nitrogen through chemical fertilizer and 25 per cent nitrogen through compost + 25 per cent nitrogen through *Azotobacter* + use of zinc @ 5 kg/ha.
Ginger (*Zingiber officinure* Rosc.)	Use of Azofert @ 20 kg/ha gave highest ginger yield (192 q/ha), and net returns (Rs./ha) at Palampur in silty clay loam soils having pH 6.83.
Cumin (*Cuminum cyminum*)	Use of 10-15 t/ha FYM + 30 kg N + 20 kg P/ha improved the yield and quality of cumin.
Fennel (*Foeniculum vulgare* Mill.)	Incorporation of 10 t/ha FYM along with 80 kg N + 45 kg P increased yield and soil health.
Turmeric (*Curcuma longa* L.)	20 t FYM/ha + 50 kg P_2O_5 + 50 kg N/ha + 100 kg K_2O/ha improved crop and soil health.

Chapter 6

Fruits

Integrated nutrient management practices were observed in pomegranate, nagpur mandarian, apple and walnut at Bikaner, Nagpur and Shalimar suggested use of inorganic fertilizers alongwith vermicompost, neem cake, dalweed mulch and biofertilizers like Azospirillum and Pseudomonas improved fruit yield as well as maintained soil fertillity (Table 6.1).

Table 6.1: Integrated Nutrient Management options for some Fruit Crops

Fruit Crops	INM Options
Pomegranate *(Punica granatum)*	At Bikaner (Rajasthan) use of inorganic fertilizers along with vermicompost helps in increasing yield as well as soil properties.
Nagpur mandarian *(Citrus spp.)*	At Akola (Maharashtra) in medium black soils, application of 800 g N, 300 g P_2O_5 and 600 g K_2O/plant along with neem cake @ 75 kg/tree/year increased fruit yield and sustained soil health.
Banana *(Musa paradisica)*	Any kinds of green manure, animal waste (solid/liquid), kitchen waste, crop residues or even high C: N ratio material such as coir dust/saw dust/rice husks can be used as a basal organic material/amendment for banana crop. If material/amendment used have high C: N ratio, it is advisable to use low C: N ratio materials like poultry manure, cattle manure *etc.* to achieve quick decomposition in situ. It is a shallow rooted crop (45 – 60 cm). Therefore, application of any kind of easily decomposable materials as a top dressing at 3 months interval is highly beneficial.
Apple *(Malus domestica* Borkh)	In apple, highest fruit yield was obtained when antitranspirant (Salicylic acid) was combined with Dal-weed mulch.
Walnut *(Juglans regia* L.)	The application of *Azospirillum* + 10 g N plant[-1] and *Psedomonas* + 5 g phosphorus and leaf area as compaired to only N @ 15 g/plant and only P 5g/plant only respectively at Shalimar, J&K.
Pine apple	A suitable and freely available organic amendment must be applied as a basal dressing into the planting furrows *e.g.* plant residues and animal waste. Foliar application of liquid fertilizer during dry period is more beneficial. Crop residues from previous pineapple crop could be composted and utilized for the next crop.

Chapter 7

Ornamental Plants

Flowers are very important part of a human life from worship to aesthetic value. Flowers are grown in almost every states of our country according to climate and needs. Use of integrated nutrient management practices in gladiolus, rose, jasmine and marigold at Allahabad, Shalimar and Bangalore increased the flower yield (Table 7.1). It suggested that the INM is as equally important to flower plants as in other cereals, pulses and vegetables.

Table 7.1: Integrated Nutrient Management Options for Ornamental plants

Regions/Location Soil Type/Soil Order Ecosystem/ Agro-climatic Regions	INM Options
Gladiolus *(Gladiolus hybrida* L.) Allahabad (U.P.)	The combined application of organic manures and inorganic fertilizers *i.e.* 75 per cent RDF + FYM + VCM + VAM resulted in the higher plant growth, plant height number of leaves/plant, increased number of florets/spike, improved flowering and increased length of spike.
Rose *(Rasa sps.)* Karnataka, Jammu and Kashmir Sub humid and semi-arid	A marginal influence in flower yield was recorded with the use of *Azotobacter* as compared to control. Uses of biofertilizer cannot be ignored since much chemical fertilizers were saved besides the yield was at par with control.
Jasmine *(Jasminum sps.)* Karnataka, Kashmir Arid, semi-arid	A use of biofertilizer coupled with organic manure (FYM, crop residues) proved beneficial and increase the flower yield by 26 percent over control.

Regions/Location *Soil Type/Soil Order* *Ecosystem/* *Agro-climatic Regions*	*INM Options*
Marigold *(Tagetus sps.)* Kashmir Silty clay loam Semi-arid	*Azotobactor* and PSM biofertilizers were tested in Kashmir to observe the potentialities to bear flowers in marigold and result recorded, indicate the 17 percent increased flower yield over control. Use of biofertilizers sustained flower productivity as well as soil health.

Chapter 8

Miscellaneous Crops

The INM practices were also observed in aswangandha, vetiver, foxtail millet, sesame, linseed and potato at various places of their domain are given in Table 8.1.

Table 8.1: Integrated Nutrient Management Options for some Miscellaneous Crops

Miscellaneous Crops	INM Options
Aswangandha	In black soils of indore incorporation of FYM @ 2.5 t/ha along with application of 12.5 kg N and 25 kg P_2O_5/ha increased the aswangandha yield.
Vetiver (Veteveria ziaanioides L.)	In shallow black soils of Malwa (M.P.) application of N (75) + P_2O_5 (40) kg/ha along with Rhizobium @ 2 kg/ha resulted in increased net profit and B: C ratio.
Foxtail Millet (Eleusine oracona Gaertn)	At Bangalore in sandy loam soils, application of N (22.5 kg) + P_2O_5 (11.25 kg) + enriched FYM @ 7.5 t/ha increased yield and foxtail millet.
Sesame (Sesamum indicum Linn.)	At Jalgaon (Maharashtra) in vertisols, application of 50 per cent recommended N through urea + 50 per cent N through pressmud/cotton cake/castorcake/compost along with recommended P and K increased crop yield and sustained soil health.
Linseed (Linum usitatissimum Linn.)	At Kota (Rajasthan) application of N: P: Zn: S (90+30+5+50 kg/ha) along with inoculation of biofertilizer (Azotobacter + PSB) and FYM @ 9 t/ha increased linseed yield and soil health.
Potato (Solanum tuberosum)	Highes tuber yield in potato (250.57 q/ha) was obtained with 75 per cent NP + seed treatment with urea + sodium bicarbonate + biofertilizers at Shalimar, J&K.
Tea (Camellia sp.)	The use of organic manures in the form of digested coirpith compost (DCC) and biofertilizers like Azospirillum brasilense, Vesicular Arbuscular Mycorrhizae (VAM) and phosphobacteria increased nutrient content and certain enzyme activities in tea hence increased productivity.

Chapter 9

INM Options in some Cropping Systems

9.1 Rice-Wheat Cropping System

Rice and wheat grown sequentially in an annual rotation constitute a rice-wheat cropping system. In annual cycle suitable thermal conditions for both rice and wheat exist in warm-temperate and subtropical areas and high altitudes in the tropics. Total contribution of rice-wheat cropping system in total cereal production is 85 percent. More than 10 M ha area is occupied by system in India. It is a dominant cropping system in the Indo-Gangetic and Non-Indo Gangetic plains in India. Major growing states are Punjab, Haryana, U.P., Uttranchal, Bihar, West Bengal and Himachal Pradesh. Rice is commonly transplanted into puddled soils and prefers continued submergence, wheat is grown in upland well-drained soils having good tilth. The INM practices in rice-wheat cropping system noticed at various locations are given in Table 9.1. The soil test based recommendations of nutrients coupled with organics and biofertilizers proved beneficial for yield as well as soil health. Dhaencha *(Sesbania aculeate)* and *Sesbania rostrate* are good sources of green manures that can be grown in rice-wheat system. GM crops are planted during leg period of 60-75 days between harvest of wheat crop and transplanting of rice after 10 July.

Table 9.1: Integrated Nutrient Management Options for Rice-Wheat Cropping System

Regions/Location Soil Type/Soil Order Ecosystem/ Agro-climatic Regions	INM Options
Ludhiana (Punjab) Loamy sand soil semi-arid	75 per cent NPK alongwith 25 per cent N supply by Jantar green manure to rice crop and 100 per cent NPK for wheat crop sustained crop yield and soil health.

Regions/Location Soil Type/Soil Order Ecosystem/ Agro-climatic Regions	INM Options
Hoshiyarpur and Kapurthala (Punjab) Loamy soil and clay loam Sub-humid	Conjoint use of 75 per cent NPK along with 25 per cent green manure or FYM @ 6 t/ha for rice and 75 per cent NPK to wheat found at par with 100 per cent NPK to rice and wheat. It indicate a net saving of 25 per cent NPK by using green manure and FYM in rice-wheat cropping system.
Jabalpur (M.P) Vertisols Sub-humid	Combine application of either 50 per cent recommended dose of NPK or 100 per cent recommended dose of NPK along with green manure for rainy season rice and 100 per cent recommended dose of NPK along with FYM 6 t/ha for wheat proved beneficial in rice-wheat cropping system.
Sambalpur (Orrisa) Sandy clay loam Sub-humid	Incorporation of green manure and 50 per cent recommended dose of NPK to rice crop and 100 per cent NPK to wheat crop improved soil health and crop yield.
Meerut (U.P) Sandy loam Semi-arid	Combined use of fertilizer @ 75 per cent NPK + 25 per cent N as FYM/ PSM to rice and 100 per cent NPK to wheat crop.
J&K, H.P. Alluvial, sandy loam	Application of 40 kg N + FYM/G.M. @ 15 t/ha + 20 kg $ZnSO_4$ to rice and 120 kg N + 80 kg P_2O_5 (through S.S.P.) + 40 kg K_2O to wheat crop improved productivity of rice-wheat cropping system and sustained soil health.
North Bengal, Assam Soils are alluvial, red and brown hill with acidic reaction	Combine use of chemical fertilizer (40 kg N+20 kg P_2O_5 + 40 kg K_2O) + FYM @ 5 t/ha/GM + Azolla @ 10 t/ha + 20 kg $ZnSO_4$ once in 3 years _ 5 kg borax to rice and 50 kg N + 20 kg P_2O_5 + FYM @ 5 t/ha to wheat crop found beneficial.
Plains of Bengal Red and laterite	Use of 40 kg N + 45 kg P_2O_5 + 30 kg K_2O + FYM/GM @ 10 t/ha + Azolla @ 10 t/ha/BGA @ 10 kg/ha + 20 kg $ZnSO_4$ to rice crop and 90 kg N+45 kg P_2O_5 + 45 kg K_2O to wheat crop.
Bihar, Eastern U.P. Alluvial red and laterite	Integrated use of fertilizers (40 kg N + 30 kg P_2O_5 + 20 kg K_2O) along with GM (green gram stover) + 20 kg $ZnSO_4$ in calcareous soil to rice and 90: 60: 30 (N+P_2O_5+K) + FYM @ 10 t/ha to wheat.
Western U.P. Alluvial	Integration of fertilizers (90 kg N + 30 kg K_2O) + FYM/GM (Sesbania/ Leucaena ropping) @ 10 t/ha to rice crop and 90 kg N + 60 kg P_2O_5 + 30 kg K_2O to wheat crop.
Lucknow (U.P) Alluvial	Integrated management of green manuring with *Sesbania rostrata* and application of BGA @ 12.5 kg soil based inoculum and chemical fertilizer has emerged a promising packages for nitrogen management for yield optimization in rice and 100 per cent NPK to wheat crop improved productivity of rice-wheat cropping system and sustained soil health
Hisar (Haryana) Sandy loam Arid	Soil test based fertilizers dose + 7.5 t FYM/ha or 15 t FYM/ha to rice-wheat.

9.2 Rice-Rice

When rice is grown after rice annually constitude a rice-rice cropping system which is followed in many states like Assam, West Bengal *etc*. The rice straw recycling in such cropping was observed at several places. The rice straw left behind after the previous crop is added to the next crop at recommended rates. A 12 – 15 per cent increases in yield was recorded over control.

About 25 to 50 per cent NPK can be substituted by incorporation of *sesbania* green leaves, organic manures act as a slow releasing fertilizers and prevent the nitrogen losses through leaching and volatilization when used in conjunction with mineral fertilizers. A few examples of INM practices are given in Table 9.2.

Table 9.2: Integrated Nutrient Management Options for Rice-Rice Cropping System

Regions/Location Soil Type/Soil Order Ecosystem/ Agro-climatic Regions	INM Options
Assam, Bengal Alluvial, red and brown hill with acidic reaction	Combine use of chemical fertilizer @ 20 kg N + 20 kg P_2O_5 + 15 kg K_2O + FYM/GM @ 10 t/ha + Azolla @ 10 t/ha + 20 kg $ZnSO_4$ once in 3 years and 60 kg N + 40 kg P_2O_5 + Azolla @ 10 t/ha in rice-rice system increase and productivity and sustained soil health.
Plains of Bengal Alluvial and red soil	Application of 60 kg N + 40 kg P_2O_5 + 30 kg K_2O + FYM/GM @ 10 t/ha + 20 kg zinc sulphate and 90 kg N + 80 kg P_2O_5 + 60 kg K_2O + Azolla @ 10 kg/ha.
Coastal plains and Ghats and Deccan Plateau Lateritic, alluvial and black	Incorporation of FYM/GM @ 5 t/ha coupled with 75 kg N + 15 kg P2O5 and 15 kg K2O and Azolla @ 10 t/ha + BGA @ 10 t/ha + 90: 60: 40 NPK improved yield and soil health.
Maruteru (A.P.) Clay loam Semi-arid	50 per cent recommended NPK through fertilizers + 50 per cent through seabania green leaf in rainy season and 100 per cent RDF of NPK through fertilizers.
Jorhat (Assam) Alluvial, sandy loam	INM packages consisting organic manure @ 1-3 t/ha (dry weight basis) + *Azospirillum* + PSB + P_2O_5 @ 10 kg/ha as rock phosphate and K_2O @ 10-20 kg/ha considered as a suitable alternative for nutrient management in rice-rice system.
24 parganas (W.Bengal)	Depending upon organic matter and nitrogen status of soil, *Azolla* and BGA biofertilizers, either alone or in combination can be used to optimise rice yield in both kharif and summer seasons at 20-25 per cent less use of N-fertilizers from that of recommended in alluvial and laterite soils of West Bengal.
Pondicherry Clay loam	In kharif rice-summer rice combine use of BGA + *Azospirillum* in combination for N nutrition in rice might substitute 20-25 per cent chemicl fertilizers N use from recommended (80-100 kg/ha) without any loss in yield. Yield optimization response of biofertilizers was less in soils showed higher response to high dose of N as chemical fertilizer.
Kharagpur (W.Bengal) Sandy clay loam Sub-humid	75 per cent NPK + 25 per cent FYM to rainy season and 100 per cent NPK to winter season improved soil fertility and plant nutritions.

9.3 and 9.4 Maize-Wheat and Soybean-Wheat

Integrated use of fertilizers along with green manuring or lantana mulch helps in improving the yield of these cropping systems besides improving soil health. Nutrient management practices followed at various locations in maize-wheat and soybean-wheat cropping systems are given in Tables 9.3 and 9.4. For optimum benefit of INM practices, it is advisable to use micronutrients fertilizers like $ZnSO_4$ @ 25 kg/ha to any component crop in soybean-wheat system in alternate years on Zn deficient soils.

Table 9.3: Integrated Nutrient Management Options for Maize-Wheat Cropping System

Regions/Location Soil Type/Soil Order Ecosystem/ Agro-climatic Regions	INM Options
Jammu and Kashmir, H.P. and Uttranchal Alluvial, brown hilly and sandy soil	Integrated use of fertilizers (60 kg N + 30 kg P_2O_5 + 20 kg K_2O) along with 10 t/ha FYM + fresh Eupatorium/Lantana mulch @ 10 t/ha to maize and 80 kg N + 30 kg P_2O_5 + 15 kg K_2O to wheat crop increased yield and soil health in maize-wheat system.
Bihar, Eastern U.P Alluvial with neutral alkaline reaction soil	Combine application 60 kg N + 60 kg P_2O_5 + GM + 16 kg borax in calcareous soil to maize and 90 kg N + 60 kg P_2O_5 + 30 kg K_2O with FYM @ 10 t/ha proved effective.
Western U.P. Alluvial	Application of 50 kg N + 20 kg K_2O/ha + FYM @ 10 t/ha to maize and 120 kg N + 60 kg P_2O_5 + 40 kg K_2O to wheat crop sustained soil health and crop yield.
Himachal Pradesh Brown hill soil	Application of FYM to meet 50 per cent of recommended N helps in economizing 50 per cent of recommended NPK in maize under maize-wheat system. Maize receiving FYM requires only half of the recommended dose of fertilizers to produce yields equal to full RDF dose applied as fertilizer.

Table 9.4: Integrated Nutrient Management Options for Soybean-Wheat Cropping System

Regions/Location Soil Type/Soil Order Ecosystem/ Agro-climatic Regions	INM Options
Rajasthan, M.P. Alluvial, red and black soils	Application of 10 kg N + 25 kg P_2O_5 (through boronated SSP) + FYM @ 4 t/ha + *Rhizobium* + 25 kg zinc sulphate in alternative year to soybean and 90 kg N + 45 kg P_2O_5 to wheat crop.
Maharashtra and Southern M.P.	In soybean crop 10 kg N + 25 kg P_2O_5 + 4 t/ha FYM and 90 kg N + 45 kg P_2O_5 to wheat crop help in sustaining soil health and crop yield.
Indore (M.P.) Vertisols Semi arid	Substitute 50 per cent of fertlizers N through 10 t FYM/ha in soybean-wheat system.

Regions/Location Soil Type/Soil Order Ecosystem/ Agro-climatic Regions	INM Options
Bhopal (M.P.) Vertisols Semi-arid	Application of 8 t/ha FYM to soybean and 60 kg N + 11 kg P/ha to wheat or application of 4 t FYM/ha + 10 kg N + 11 kg P/ha to soybean and 90 kg N + 22 kg P/ha to wheat gave the yield of 2 t/ha soybean and 3.5 t/ha wheat besides sustaining soil health.

9.5 Miscellaneous Cropping Systems

INM options have been recorded in rice-mustard, rice-potato, sugarcan-potato, pearlmillet-wheat, sorghum-gram, groundnut-wheat, sugarcane-wheat, rice-gram, sorghum-chickpea, soybean-winter maize, sugarcane-wheat, sugarcane-lentil, groundnut-pearlmillet and summer mungbean-winter wheat at various locations of the country focussed the combined use of mineral and organic plant nutrient sources are given in Table 9.5. A single source of plant nutrients neither fulfilled the all needs of a crop nor provides balance nutrition, hence all sources should be used judiciously accordingly to the need, availability and climatic condition in an integrated manner.

Table 9.5: Integrated Nutrient Management Options for Miscellaneous Cropping System

Regions/Location Soil Type/Soil Order Ecosystem/ Agro-climatic Regions	Cropping System	INM Options
H.P., J&K, Uttranchal Alluvial, sandy loam	Rice- Mustard	Use of 40 kg N + 30 kg P_2O_5 + 40 kg K_2O + FYM/GM @ 10 t/ha + *Azolla* @ 10 t/ha + 20 kg $ZnSO_4$ to rice crop and 20 kg N + 10 kg P_2O_5 + 25 kg K_2O to mustard crop.
-do-	Rice-potato	Applications of 40 kg N + 20 kg P_2O_5 + 15 kg K_2O + *Azolla*/GM @ 10 t/ha + 20 kg $ZnSO_4$ to rice and 50 kg N + 50 kg P_2O_5 + 30 kg K_2O + FYM @ 10 t/ha + seed treatment with *Azotobacter* and PSB to potato crop gave highest yield.
Western U.P. Alluvial with neutral alkaline reactions	Sugarcane-potato	In autumn planted sugarcane, use of 100 kg N + 45 kg P_2O_5 + 30 kg sulphitation pressmud/GM + incorporation of potato foliage. In potato, application of 135 kg N + 20 kg P_2O_5 + 60 kg K_2O + FYM @ 10 t/ha + seed treatment with *Azotobacter* + PSB.
S.K.Nagar (Gujarat) Sandy loam Arid	Pearlmillet-wheat	50 per cent NPK + 50 per cent organic manure (FYM) to rainy season pearlmillet and 100 per cent fertilizer to winter wheat.

Regions/Location Soil Type/Soil Order Ecosystem/ Agro-climatic Regions	Cropping System	INM Options
Hisar (Haryana) Sandy loam Arid	Pearlmillet-wheat	75 per cent RDF through fertilizer + 25 per cent N through FYM to kharif crop and 75 per cent RDF through fertilizer to winter wheat.
Modipuram (U.P.) Sandy clay Semi-arid	Sorghum-gram	Incorporation of FYM @ 6 t/ha + 50 per cent RDF to sorghum as well gram proved effective.
Kanpur (U.P.) Sandy loam Semi-arid	Groundnut-wheat	20: 30: 45 (N: P: K) + FYM @ 10 t/ha + gypsum @ 300 kg/ha to groundnut and 80: 40 (N: P) to wheat crop.
Sangli (Maharashtra) Medium black Semi-arid	Groundnut-wheat	Incorporation of 2.5 t FYM + 75 per cent recommended fertilizers to kharif groundnut and 100 per cent NPK to wheat crop.
-do-	Sugarcane-wheat	In sugarcane 135 kg N + 45 kg P_2O_5 + 30 kg K_2O + (FYM + sulphitation pressmud)/GM (Sesbania/sunhemp) @ 10 t/ha and 80 kg N + 40 kg P_2O_5 + 40 kg K_2O to wheat sustained soil and crop productivity.
Rajasthan, M.P. Alluvial, red and black soils	Rice-gram	Application of 25 kg N + 15 kg K_2O + pulse crop residues incorporation + BGA @ 10 kg/ha/*Azolla* @ 10 t/ha to rice and 10 kg N + 20 kg P_2O_5 + *Rhizobium* + 5 t FYM/ha + 500 g PSB to gram improved yield and soil health.
Coast plains and Ghats soils are lateritic black and coastal alluvial	Pearlmillet-wheat	Combine use of 25 kg N + 20 kg P_2O_5 + 10 kg K_2O + *Azotobacter/Azospirillum* to pearlmillet and 45 kg N + 20 kg P_2O_5 + 15 kg K_2O + FYM @ 5 t/ha to wheat increased yield.
Udaipur (Rajasthan) Sandy loam Semi-arid	Sorghum-chickpea	Combined application of *Azospirillum* + *Azotobacter* and PSB biofertilizers with 5 t/ha organic manure at 20 and 50 per cent reduced level of N and P supply from recommended (80 kg N and 40 kg P_2O_5/ha) can provide at least 25 per cent higher grain yield of sorghum that achieved with 100 per cent RDF.
		In chickpea application of *Rhizobium* + PSB and 5 t organic manure/ha at 50 per cent reduced level of N and P fertilizer application from the recommended (20 kg N and 40 kg P_2O_5/ha) can provide 60 per cent higher grain yield of chickpea from achieved with 100 per cent RDF.
Udaipur (Rajasthan) Sandy loam Semi-arid	Soybean-winter maize	Combined application of Rhizobium + PSB + 5 t organic manure/ha @ 80 per cent RD of N (20 kg/ha) and 50 per cent of RD of P (40 kg/ha P_2O_5) in soybean and *Azospirillum* + *Azotobacter* and PSB with 5 t/ha organic manure at 80 per cent of the recommended N dose (150 kg/ha) and 50 per cent of the recommended P dose (160 kg P_2O_5/ha) was found most promising for recommendation.

Regions/Location Soil Type/Soil Order Ecosystem/ Agro-climatic Regions	Cropping System	INM Options
Pantnagar Medium alkaline soil (Uttranchal)	Sugarcane-sugarcane-wheat/Lentil/Mentha	Application of combination of *Azotobacter diasotrophicus*, *Azospirillum*, compensate can yield due to 30 – 40 per cent less application of fertilizer nitrogen (150 kg N/ha) + recommended nutrient of K_2O and P_2O_5.
Saurashtra (Gujarat)	Groundnut-pearlmillet	Use of PSM strain at 50 per cent reduced level of P as DAP (50 kg P_2O_5/ha for ground nut and 40 kg P_2O_5/ha for pearlmillet) shall have similar yield that obtained under 100 per cent recommended dose of P fertilizer.
Vidarpha (MS) Clay loam soil	Summer mungbean-winter wheat-pearlmillet-chickpea	Organic manure @ 5 t/ha + 20 kg N + 40 kg P_2O_5/ha and bradyrhizobium + PSB appeared to the best package for mungbean and wheat. Wheat yield was satisfactory at par with that of 75 per cent RDF (100 kg N + 60 kg P_2O_5/ha) best package
Delhi Alluvial, Medium alkaline	Rice-wheat-mungbean	Combined application of BGA and *Azolla* @ 60 kg/ha N supply from chemical fertilizer improve the nitrogen economy of kharif HYV rice in the specified soil locations provides yield at par with 90-120 kg N supply/ha as chemical fertilizer. Combined application of BGA + PSB reduced 30 kg P_2O_5/ha applied as single super phosphate in the recommended schedule of chemical fertilizers without any yield loss.

References

Abusaleha and Shanmugavelu, K.G. (1988). Effect of organic vs inorganic sources of nitrogen on growth, yield and quality of okra. *Indian Journal of Horticulture* 29: 312-318.

Acharya, C.L., Subbarao, A. Biswas, A.K., Sammireddy, K., Yadav, R.L., Divedi,B.S., Shukla, A.K., Singh, V.K. and Sharma, S.K. (2003). Methodoloies and package of practices on improved fertilizer use efficiency under various agroclimatic regeions for different crops/cropping systems and soil conditions Indian Institute of Soil Science, Bhopal – 38 pp. 1-74.

Alam, A. (2003). Annual Report, University of Agricultural Sciences and Technology of Kashmir, Shalimar, Srinagar, 191 121, India.

Alam, A. and Wani, S.A. (2003). Emerging need for organic agriculture and strategies for its optimization. In souvenir of National Seminar on organic products and their future prospects, organized at SKUAST-K, Shalimar, Srinagar.

Anonymous. (2002). The Hindu Survey of Indian Agriculture.

Ariyarathe (2000). Integrated Plant Nutrition System (IPNS). Training Manual (Srilanka).

Balanced fertilizer use for crops on http: 11www.indiaagronet.com/indiaagronet/ Technology – Upd/contents/integrated-nutrient. 10/30/2003.

Bhardwaj, K.K.R. (1998). Role of organic waste in soil fertility management In: 50 years of Natural Resource Management Research (G.B. Singh and Sharma, B.R., Eds.) NRM Division, ICAR, New Delhi pp. 213-220.

Bhattachary, P. and Mishra, U.C. (1995). A book on biofertilizer for extension workers published by the National Biofertilizer Development Centre, Ghaziabad.

Doran, J.W., Jones, A.J., Arshad, M.A. and Gilley, J.E. (1999). Determinants of soil quality and health. In: soil quality and health. In soil quality and soil erosion (R. Lal Ed.), CRC Press, New York, USA p 17-36.

Easwaran, S., Kumar, N. and Marimuthu, S. (2006) Studies on the effect of integrated nutrient management on leaf nutrient status and enzymes activities in tea (Camellia sp.). Indian Journal of Horticulture 63: 224-226.

Gicana, R. Norlito. (2001). Integrated plant nutrient management in the Phillipines. In proceedings of Regional Workshop on IPNS development in rural poverty alleviation, Bangkok, Thailand.

Harmson, K. (1995). Integrated phosphorus management. In integrated plant nutrition systems. FAO fertilizer and plant nutrition bulletin 12: pp 293-306.

Hegde, D.M. and Dwivedi, B.S. (1993). Integrated nutrient supply and management as a strategy to meet nutrient demand. Fertilizer News 38: 49-59.

Integrated nutrient management: Role of biofertilizer, Recommendations of DBT network programme-1999-2003 Department of Biotechnology, Ministry of Science and Technology, Government of India pp 1-18.

Integrated plant nutrition system (IPNS) compendium (2002). Published by Economic and social commission for Asia and the pacific, Fertilizer Advisory, Development and Information Network for Asia and the Pacific (FADINAP) pp 1-53.

Jose, D., Shanmugavelu, K.G. and Thamuraj, S. (1988). Studies on the efficacy of organic vs inorganic form of nitrogen in brinjal. Indian Journal of Horticulture 27: 100-103.

Lampe, Siegfried. (1999). Principles of Integrated plant Nutrition Management System. FADINAP | FPA workshop on development of integrated plant nutrition system for the Philippines, Ternate, Carite, Phillippines.

Marwaha, B.C. (1995). Biofetilizers-A supplementary source of plant nutrient. Fertilizer News 40: 39-50.

Marwaha, B.C. (2003). Scope and significance of fertigation in Indian Agriculture. Fertilizer News 48: 97-101.

Motsara, M.R. and Bisoyi, R.N.(2001). 50 crop demonstrations on biofertilizers. National Biofertilizer Development Centre, Ghaziabad (U.P.) India.

Motsara, M.R., Bhattacharayya, P. and Srivastava Beena (1995). Biofertilizer Technology, Marketing and Usage, A source Book Cum Glossary,FDCO Publication, New Delhi.

Nath, B. and Korla, B.N. (2000) Studies on effect of biofertilizers in ginger. Indian Journal of Horticulture 57: 168-171.

NBSSLUP (2004). Soil Resource Management Report. National Bureau of Soil Survey and Land Use Planning, Nagpur.

Parr, J.F., Papendick, R.I., Hornich, S.B. and Meyer, R.E. (1992). Soil quality: attributes and relationship to alternative and sustainable agriculture. American Journal Alternative Agriculture 7: 5-11.

Raghavaiah, C.V., Verma, V.S., More, S.D., Patel, M.K. and Hebbara, M. (2003). Oilseed production in salt affected soils. P 40.

Sharma, P.D. and Biswas, P.P. (2004). IPNS packages for dominant cropping system in different agro-climatic regions of the country. Fertilizer News 49: pp 43-47.

Sharma, R.P. and Rana, D.S. (1993). Nutrient Management in vegetable crops for sustainable production. Fertilizer News 38: 31-44.

Singh, A.K., Narayansamy, D., Rattan, R.K., Singh, D. Shachdev, P., Mukhopadhay, S.S. (2001). Abstracts, National Seminar on Development in Soil Science, IISS, New Delhi, p 245.

Singh, A.K. (2006) Mantra for the new millennium. The Hindu Survey of Indian Agriculture pp 32-40.

Singh, K., Gill, I.S. and Verma, D.P. (1970). Studies on poultry manure in relation to vegetable production cauliflower. Indian Journal of Horticulture 20: 537-547.

Singh, K., Minhas, M.S. and Srivastava, D.P. (1969). Studies on poultry manure in relation to vegetable production (potato). Indian Journal of Agricultural Research 26: 69-73.

Singh, S.R., Sant, P. and Kumar, J. (2000). Organic farming technologies for sustainable vegetable production in Himachal Pradesh. Himachal Journal of Agriculture Research 26: 69-73.

Somani, L.L. and Totawat, K.L. (2001). Integrated nutrient management for maintaining soil and crop productivity in. Souvenir of the 66th annual convention of the Indian Society of Soil Science (Somani, L.L. and Totawat, K.L. eds.) organized by Department of Agricultural Chemistry and Soil Science, Rajasthan College of Agriculture, Udaipur.

Srinivasarao, Ch., Prasad, J.V.N.S., Vittal, K.P.R., Venkateswar W.B. and Sharma, K.L.C. (2003) Role of optimum plant nutrition in drought management in rainfed agriculture, Fertilizer News pp 105-114.

Subbarao, A., Subhash Chand, Srivastava, S. (2002). Opportunities for integrated plant nutrient supply system for crops/cropping system in different agro-eco-regions. Fertilizer News, 47: 75-90.

Subhash Chand (2001) Integrated nutrient management in mustard (Braccica juncea L. (Zern and Coss.) Ph.D. thesis, MPUAT, Udaipur, Rajasthan.

Subhash Chand (2007). Dictionary of Soil Science, Daya Publisher, New Delhi pp 1-126.

Subhash Chand and Mir, A.H. (2007). Micro-organisms as a component for improving phosphate use efficiency by crops. Journal of Environment and Ecology, 23 pp 186-189.

Subhash Chand and Pabbi, S. (2005). Organic farming – a rising concept. In souvenir of Agriculture summit – 2005 organised by Ministry of Agriculture, Government of India and FICCI, Vigyan Bhavan, New Delhi pp. 1-8.

Subhash Chand and Somani, L.L. (2003). Balanced use of fertilizer, organics and biofertilizer for improving yield of mustard. *International Journal of Tropical Agriculture*, 21: 133-140.

Subhash Chand and Somani, L.L. (2005). Exploring possibilities of improving the yield of mustard (*Brassica juncea* (L.) (Czern and Coss) through integrated nutrient management. *International Journal of Tropical Agriculture* 23: 177-182.

Subhash Chand, Sahi, N.C. and Ali, T. (2006). Vermicomposting in organic farming. In souvenir of Agriculture summit-2006 organised by Ministry of Agriculture, Govt. of India and FICCI, Vigyan Bhavan, New Delhi pp. 1-4.

Tiwari, K.N., Sulewski Garin and Portch Sam (2005). Challenges of meeting nutrient needs in organic farming. *Indian Journal of Fertilizers* 4: 41-48; 51-59.

Tondon, H.L.S. (1996). In: Plant Nutrient Needs, Supply, efficiency and policy issues. (J.S. Kanwar and J.C. Katyal, Eds.) pp. 15-28. National Academy of Agricultural Sciences, New Delhi, India.

Wani, S.A., Dar,G.H., Singh, G. and Sofi, N.A. (2003). Souvenir of the national seminar on organic products and their future prospectus. SKUAST-K, Shalimar Campus, Srinagar.

Whitbread, M. Antony, Blair, J. Graeme, and Letroy,D.B. Rod. (2000). Managing legume leys, residues and fertilizers to enhance the sustainability of wheat cropping system in Australia 2. Soil physical fertility and carbon. *Soil and Tillage Research* 54: 57-89.

Yadav, D.S. and Alok Kumar.(1993). System based integrated nutrient management for sustainable crop production in Uttar Pradesh. *Fertilizer News* 38: 45-51.

Yadav, R.L. (2006). Improving nitrogen efficiency. The Hindu Survey of Indian Agriculture p. 108.

Appendices

Appendix-1: Nutrient Essential for Plant Growth and Forms in which Nutrients are taken up by Plants

Nutrient	Chemical Symbol	Form taken up by Plant
Primary Nutrients		
Carbon	C	CO_2, HCO_3
Hydrogen	H	H_2O
Oxygen	O	H_2O, O_2
Nitrogen	N	NH_4^+, NO_3^-
Phosphorus	P	$H_2PO_4^-$, HPO_4^{-2}
Potassium	K	K^+
Secondary Nutrients		
Calcium	Ca	Ca^{2+}
Magnesium	Mg	Mg^{2+}
Sulphur	S	SO_4^{2-}
Micro-nutrients		
Iron	Fe	Fe^{2+}, Fe^{2+}, Chelate
Zinc	Zn	Zn^{2+}, $Zn(OH)_2^0$, chelate
Manganese	Mn	Mn^{2+}, chelate
Copper	Cu	Cu^{2+}, chelate
Boron	B	$B(OH)_3^0$
Molybdenum	Mo	MoO_4^-
Chlorine	Cl	Cl^-

Appendix-2: Keys to Nutrient Deficiency Symptoms in Crops

Nutrient	Colour Change in Lower Leaves
N	Plant light green, older leaves yellow
P	Plants dark green with purple cast, leaves and plants small
K	Yellowing and scorching along the margin of older leaves
Mg	Older leaves have yellow discolouration between veins-finally reddish purple from edge inward.
Zn	Pronounded interveinal chlorosis and bronzing of leaves
Nutrient	*Colour Change in Upper Leaves (Terminal bud dies)*
Ca	Delay in emergence of primary leaves, terminal buds deteriorate
B	Leaves near growing point turn yellow, growth buds appear as white or light brown, with dead tissue
Nutrient	*Colour Change in Upper Leaves (Terminal bud remain alive)*
S	Leaves including veins turn pale green to yellow, first appearance in young leaves.
Fe	Leaves yellow to almost white, interveinal chlorosis at leaf tip
Mn	Leaves yellowish gray or reddish, gray with green veins
Cu	Young leaves uniformly pale yellow. May wilt or wither without chlorosis
Mo	Wilting of upper leaves, then chlorosis
Cl	Young leaves wilt and die along margin

Appendix-3: List of Biofertilizers, their Functions and Beneficiaries Crops

Sl.No.	Name of the biofertilizer	Function/Contribution	Beneficiaries (Crops)
		A. Nitrogen Biofertilizers	
1.	Rhizobium (symbiotic)	1. Fixes 50-100 kg N/ha 2. Increase yield from 10-35 per cent 3. Leaves residual nitrogen	Pulse legumes, Oilseed legumes, Fodder legumes, Forest legumes
2.	Azotobacter (non-symbiotic)	1. Fixation of 20-25 kg N/ha 2. 10-15 per cent increase in yield 3. Production of growth promoting substances	Wheat, maize, cotton, sorghum, sugarcane, pear, millet, rice, vegetables and several other crops
3.	Azospirillum (associative)	1. Fixation of 20-25 kg N/ha 2. 10-15 per cent increase in yield. 3. Production of growth promoting substances	Wheat, maize, cotton, sorghum, sugarcane, pear, millet, rice, vegetables and several other crops
4.	Blue green algae (BGA) (phototropic)	1. Fixation of 20-30 kg N/ha 2. 10-15 per cent increase in yield 3. Production of growth promoting substances	Flooded rice.
5.	Azolla (symbiotic)	1. Fixation of 30-100 kg N/ha 2. 10-15 per cent increase in yield	Only for flooded rice
		B. Phosphorus Biofertilizers	
1.	Phosphate Solubilizing Microorganisms (Bacteria/fungi)	1. Solubilizers inso-luble phosphates. 2. Yield increase from 10-20 per cent	All types of crops
2.	VAM (Obligate symbionts)	1. Enhance uptake of P, Zn, S, Fe, Cu and water 2. Promotes uniform crop, increase growth and yield	Forest trees

Appendix-4: Average Nutrient Composition of Organic Material (Oven-dry basis)

Kind of Material	Total N	Total P_2O_5	Total K_2O	Total CaO
Compost	1.34	3.30	1.04	0.89
Swine manure	0.81	3.00	0.61	4.75
Carabao manure	0.60	2.05	0.50	-
Cow manure	1.87	2.47	2.11	-
Goat manure	2.81	2.66	1.20	-
Horse manure	3.13	2.80	1.88	-
Sludge	1.87	3.11	0.54	4.30
Vermi-cast	1.86	3.61	1.60	2.21
Azolla	2.76	0.97	2.38	1.09
Rice straw	0.48	0.34	1.58	-
Coconut coir dust	0.50	0.82	1.26	-
Mud press	2.72	6.20	0.79	-
Distillery slops	0.12	0.25	0.62	-
Garbage ash	0.68	T	1.40	3.45
Water hyacinth ash	0.50	8.06	19.08	-
Factory ash	0.22	2.76	0.94	0.75
Bagasse ash	0.28	0.84	2.00	-

Appendix-5: Nutrient Removed by Vegetable

Crop	Yield (t/ha)	Primary Nutrient (kg/ha)			Secondary Nutrients (kg/ha)			Micronutrients (g/ha)		
		N	P	K	Ca	Mg	Cu	Zn	Fe	Mn
Beet	52	139	20	171	29	9	90	172	2397	442
Spinach	16	58	9	88	41	17	89	62	3371	371
Lettuce	10	16	2	23	5	2	9	35	430	46
Sweet Potato (T)	23	48	13	83	10	5	26	14	416	70
Sweet Potato (Total)		108	20	207	23	15	58	38	2147	255
Radish	35	73	11	118	42	11	25	112	1295	241
Cabbage	60	128	18	175	60	14	44	188	1365	382
Bittergourd (P)	30	80	12	88	6	10	83	13	447	113
Bittergourd (Total)		130	19	148	30	21	127	150	1563	281
Luffa (P)	30	41	9	54	5	5	32	53	626	58
Luffa (Total)		151	24	169	97	33	118	218	3198	342
Snakegourd (P)	30	40	6	52	17	7	55	55	492	103
Snakegourd (Total)		73	10	89	38	14	93	96	1256	285
Bean (P)	12	39	6	39	5	4	23	60	544	131
Bean (Total)		166	19	143	86	19	106	388	3831	1606
OkrA (P)	18	59	9	76	15	10	50	87	455	158
Okra (Total)		95	14	131	62	24	86	172	1357	553
Capsicum (P)	10	25	2	14	9	5	18	20	147	44
Capsicum (Total)		67	5	63	23	9	56	70	1074	208
Brinjal (P)	25	57	9	74	1	6	47	162	153	37
Brinjal (Total)		130	18	241	58	21	141	251	571	210

Crop	Yield (t/ha)	Primary Nutrient (kg/ha)			Secondary Nutrients (kg/ha)			Micronutrients (g/ha)		
		N	P	K	Ca	Mg	Cu	Zn	Fe	Mn
Potato (T)	20	38	7	81	3	4	32	153	652	89
Potato (Total)		222	27	479	71	24	105	541	10546	2506
Tomato (F)	25	127	14	165	10	9	176	129	4103	126
Tomato (Total)		163	18	206	42	20	346	222	7709	344
Carrot (T)	31	70	14	104	23	10	53	158	1491	267
Carrot (Total)		159	24	237	52	20	114	430	3693	699

F: Fruits; P: Pods; T: Tuber.

Appendix-6: Nitrogen Content and C/N Ratio of some Compostable Materials

Materials	Nitrogen Content (Per cent)	C: N Ratio
Farm residue		
Rice straw	0.3-0.5	80-130
Wheat straw	0.3-0.5	80-130
Barley straw	0.3-0.4	100-120
Maize stalks and leaves	0.8	50-60
Cotton stalks	0.6	70
Sugarcane trash	0.3-0.4	110-120
Lucern residue	2.55	19
Green weeds	2.45	13
Water hyacinth	2.38	17.6
Seaweeds	2.10	19
Azolla	2.5	25
Red clover	1.9	44
Ferns	1.5	40.80
Flax	1.1	20
Fallen leaves	0.5-1.0	17.9
Grass clippines	2.15	4.5
Sesbania sp.	2.83	
Neem cake	6.05	
Animal shed waste		
Cow dung	2.0	
Buffalo dung		
Horsedung	2.4	
Poultry	5.0	
Sheep	3.75	
Pig	3.75	
Human habitation waste		
Night soil	4.0-6.0	6-10
Urine	15-18	0.8
Digested sludge	5.0-6.0	6
Biogas (ex-cattle) slurry	2.0	20.4

Materials	Nitrogen Content (Per cent)	C: N Ratio
Vegetable residue		
Potato tops	1.6	27
Amaranthus	3.6	11
Cabbage	3.6	12
Lettuce	3.7	15
Onion	2.6	15
Pepper	2.6	12
Tomato	3.3	27
Carrot (whole)	1.6	
Turnip top	2.3	
Fruit waste	1.5	
Tobacco	3.0	
Forest		
Leaves	0.5-1.0	40-80
Raw sawdust	0.25	208
Rotted sawdust	0.3	128
Mango sawdust	0.3	132

Appendix-7: Contribution of Rhizobium in Biological Nitrogen Fixation

Name of the Crop N. fixation	Quantum of (kg/N/ha)
1. Alfalfa	100-200
2. Redgram	166-200
3. Clover	100-200
4. Gram	85-110
5. Cowpea	80-85
6. Groundnut	50-60
7. Lentil	90-100
8. Green/Black gram	50-55
9. Pea	52-77
10. Soybean	60-80

Appendix-8: Response of Pulse Crops to Inoculation with Rhizobium Culture under different Agro-climatic Conditions

Sl.No.	Crop	Location	Increase in Grain Yield over Control (in per cent age)
1.	Mungbean	Lam (AP)	14.75
		Dholi (Bihar)	16.49
		Pantnagar (Uttarakhand)	4.15
		Delhi	13.33
2.	Urd bean	Pudukottai (TN)	4.21
		Dholi (Bihar)	11.29
		Pantnagar (Uttarakhand)	17.21
3.	Cowpea	SK.Nagar (Gujarat)	10-36
4.	Pigeonpea	Hissar (Haryana)	5-25
		Pantnagar (Uttarakhand)	2-25
		SK Nagar (Gujarat)	9-21
		Sehore (MP)	13-29
		Rahuri (Maharashtra)	3-40
5.	Chickpea	Varanasi (UP)	4-19
		Dholi (Bihar)	25-42
		Delhi	18-28
		Hissar (Haryana)	24-43
		Dohad (Gujarat)	33-67
		Sehore (MP)	20-41
		Badrapur (Maharashtra)	8-12
6.	Lentil	Pantnagar (Uttarakhand)	4-26

Appendix-9: Recommended Crops and Crop-Response to Azotobacter in different Crops

Sl.No.	Crop	Method of Inoculation	Increase over Control (percentage)
1.	Onion	Roots dipped in a slurry of legnite based inoculant at transplanting	18-122
2.	Brinjal	-do-	1-42
3.	Tomato	-do-	2-29
4.	Cabbage	-do-	25-50
5.	Wheat	Seed inocuation with cell suspension in water	10-30
6.	Sorghum	-do-	9-3-38.1
7.	Maize	-do-	36.5-71.7
8.	Cotton	-do-	6.7-26.6
9.	Rice	-do-	1-20

Appendix-10: Nutrient Requirement of different Vegetable Crops

Crop	FYM (t/ha)	N	P_2O_5 (kg/ha)	K_2O
Potato	15-30	120-150	80-100	100-120
Tomato	15-20	75-90	60-80	40-60
Brinjal	15-20	75-90	60-80	40-60
Cabbage	15-20	120-150	60-80	40-60
Califlower	15-30	100-120	50-60	50-60
Onion	15-20	80-120	50-60	30-40
Peas	10-15	20-30	50-60	30-40
Beans	10-15	20-30	50-60	30-50
Spinach	20-30	100-150	50-100	40-60
Lettuce	20-30	60-80	60-80	60-80
Celery	20-30	120-130	40-60	60-80
Amaranthus	10-15	40-60	40-60	40-60

Appendix-11: Compatibility Chart for Soluble Fertilisers

Fertilisers	Urea	AN	AS	Cal. Nitrate	MAP	MKP	Pot. Nitrate (KNO₃)	KNO₃ + Mg	KNO₃ + P	N+Mg
Urea	-	C	C	C	C	C	C	C	C	C
AN	C	-	C	C	C	C	C	C	C	C
AS	C	C	-	L	C	C	L	L	C	C
Calcium nitrate	C	C	L	-	X	X	C	X	X	C
MAP	C	C	C	X	-	C	C	L	C	X
MKP	C	C	C	X	C	-	C	L	C	X
KNO₃ₓ	C	C	L	C	C	C	-	C	C	C
KNO₃+Mg	C	C	L	X	L	L	C	-	X	C
KNO₃+P	C	C	C	X	C	C	C	X	-	X
N + Mg	C	C	C	C	X	X	C	C	X	-
Polyfeed	C	C	C	X	C	C	C	C	C	C

C: Compatible; L: Limited compatibility; X: incompatible.

Appendix-12: Key Soil and Environmental Indicators as Influenced by Agricultural Management Practices

Soil or Environmental Indicators	General trend/change	Long-term Agricultural Practices Affecting the Indicator
Soil organic matter	Increase	Continuous cropping with well-managed crop residue, zero or minimum tillage, legume-based and other crop rotations, legume plow down (green manure), cover crops, forages
	Decrease	Excessive tillage, summer fallow, crop residue removed or burned
Microbial biomass and biological diversity	Increase or decrease	Same as for soil organic matter
Soil aggregate stability	Increase	Conservation tillage, maintenance of crop residue, forages and legumes in crop rotations
	Decrease	Same as for soil organic matter decrease
Hydraulic conductivity	Increase	Reduced and zero tillage, maintenance of crop residue, forages and legumes in crop rotations-degree and extent of change vary with different practices
	Decrease	Same as for soil organic matter decrease
Soil depth/rooting volume	Increase	Conservation tillage and forage-based crop rotations should reduce erosion and allow soil-forming factors to maintain and rehabilitate top soils.
	Decrease	Excessive tillage, summer fallow cropping system, and crop residue removal or burning are the main agricultural practices that subject soils to serious wind and water erosion resulting in top soil removal
Water quality	Positive or negative	Data are lacking on how soil water quality is affected by different agricultural practices; in general, zero or minimum tillage, forage-based cropping systems, and maintenance of crop residue reduce surface runoff and soil loss to water streams; excessive use of herbicides and fertilizer may result in deterioration of water quality.

Appendix-13: Proposed Minimum Data Set of Soil Physical, Chemical and Biological Indicators for Screening the Condition, Quality and Health of Soil

Indicators of Soil Condition	Relationship to Soil Condition and Function; Rationale as a Priority Measurement
Physical	
Texture	Retention and transport of water and chemicals; needed for many process models; estimate of degree of erosion and field variability of soil types
Depth of soil, topsoil and rooting	Estimate of productivity potential and erosion, normalizes landscape and geographic variation
Soil bulk dendity and infiltration	Indicators of compaction and potentil for leaching, productivity and erosivity, density needed to adjust soil analyses to field volume basis.
Water-holding capacity (water retention character)	Related to water retention, transport, and erosivity; available water can be calculated from soil bulk density, texture, and soil organic matter
Chemical	
Soil organic matter (total organic C and N)	Defines soil fertility, stability and erosion extent; use in process models and for site normalization.
pH	Define biological and chemical activity thresholds; essential to process modeling.
Electrical conductivity	Defines plant and microbial activity thresholds, soil structural stability, and infiltration of added water, presently lacking in most process models, can be a practical estimator of soil nitrate and leach able salts.
Biological	
Microbial biomass C and N	Microbial catalytic potential and repository for C and N; modeling; early warning of management effects on organic matter
Potentially mineralizable N (anaerobic incubation)	Soil productivity and N-supplying potential; process modeling; surrogate indicator of microbial biomass N
Soil respiration, water content, and temperature	Measure of microbial activity (in some cases plants); process modeling.

Appendix-14: Taxonomic Classification of Soils of India

Soil Order	Area (Million ha)
Inceptisols	134.1
Entisols	78.7
Alfisols	42.2
Aridisols	13.3
Utisols	8.4
Mollisols	1.6
Others	23.7
Total	328.7

Appendix-15: Major Soil Groups of India

Soil Group	Area (Million ha)
Alluvial soils	93.1
Red soils	79.7
Black soils	55.1
Desert soils	26.2
Lateritic soils	17.9
Coastal alluvial	10.1
Hill soils	3.6
Tarai soils	0.3
Rock out crops	0.6
Others	42.0
Total	328.6

Appendix-16: Soil Enzymes

Name of Enzyme	Reaction Responsible
Urease	Cases hydrolysis of urea
Phosphatase	Responsible for hydrolytic cleavage of phosphate monoester
Cellulose	Hydrolysis of cellulose
Amylase	Hydrolysis of starch
Proteinase	Conversion of protein to amino acids
Sulphatase	Conversion of sulphate easter to sulphate ions

Appendix-17: Pesticides Usage on different Crops in India

Crop	Pesticide Use (per cent)	Cropped Area (per cent)
Cotton	54	05
Rice	17	24
Oilseeds	02	10
Vegetables and fruits	13	03
Sugarcane	03	02
Plantation crops	08	02
Other (wheat, coarse cereals, pulses and millets)	03	54
Total	100	100

Appendix-18: Technology for Production of Vermicompost

☆ Earthworms ingest vegetable matter, soil *etc*. and excrete small pellets of finely ground soil (called casts) very rich in nitrogen, phosphorus and potassium. They also turn the soil, thus providing air to micro-organisms and the roots of plants. Due to these properties, earthworms can be used to produce vermicompost. The attention have been increasing on breeding of earthworms (vermi-culture) and their subsequent use for preparation of vermicompost.

☆ Typically, a 1.5 hectare farm will need about 12 tonnes of biomass which can be converted to about 8 tonnes of vermicompost which would be sufficient for given area. Best results are obtained with two eregrine (exotic or foreign) species, *Eisenia foetida* and *Eudrilus eugeniae*. These rapidly convert biomass into casts and also breed very fast. Endemic species (found naturally in India), like *Perionyx excavatus* and *Perionyx sansibaricus* are also suitable but have lower conversion rates and lower breeding rates. Vermicompost units require maintenance of 40 per cent -50 per cent moisture and 20-30°C temperature consequently, regular water supply is essential.

☆ Compared to chemical fertilizers, larger quqntities of vermicompost are needed for better soil health and sustained crop production. The yields of crops are usually equivalent to those obtained by using chemical fertilizers. Typically, 1.5 tonnes of feed mixture produces 1 tonn of vermicompost in 3 months and continuous cycle has to be maintained for year-round yield of vermicompost. It takes about two or three years of regular use of vermicompost for all benefits to become apparent.

☆ Pits/Troughs can be constructed, either above or below the surface, using brick masonry or stone slabs or even plastic. For a volume of 1m3, a pit/trough of 1.6m length, 1m width and 0.75m height is suitable. The number of worms is dependent on the volume: in the above case about 6000 to 7000 worms can be used. If the pits are constructed outdoors, a thatched roof should be built over it as moisture in the 40-50 per cent range and temperature in the 20-30°C range need to be maintained.

☆ A ditch of standing water around the pit/trough and a wire mesh keeps away predators like insects, cats, dogs and birds. Any organic waste or agricultural residue can be used as feed mixture, *e.g.* farmyard wastes, green wastes, sugarcane thrash, coir and pith, kitchen wastes. These are mixed with cow-dung in the ratio 8:1 and put in the pit/trough. This feed mixture is allowed to decompose for at least 2-3 weeks. During this time, the mass will heat up. If earthworms are introduced in this period, they will die. Commercially decomposing mixture can be added to facilitate decomposition. Addition of neem powder prevents infestation. 1kg of earthworms represents 600 to 1000 worms. These can convert 45 kg of wet biomass (40 per cent moisture) in a week's time yielding about 25 kg of vermicompost.

☆ The partially decomposed biomass in the pit/trough is inoculated with earthworms. Watering is done daily. The worms feed on the biomass, assimilating 5 per cent -10 per cent for their growth and excreting the rest in the form of nutrient rich casts. Once the feed mixture is seen to largely contain casts, it is dumped in a conical heap and left for a few hours. The worms effect at the base and can be easily retrieved for reuse. The rest of the dried material is passed through a 3mm sieve to collect the casts as vermicompost. A sequential system of vermicomposting has been tried, where various stages are developed in interconnected pits and earthworms migrate from one pit to another.

Appendix-19: Technology for Production of Enriched Compost

Technology for the production of enriched compost has been recommended for field-level adoption through the Ministry of Agriculture and Cooperation. Salient features of the technology are as follows:

☆ Add 0.5 per cent mineral nitrogen and 5 per cent Mussoorie rock phosphate to the compostable material. Only compostable materials having C: N ratio higher than 50 need addition of mineral nitrogen. Composting of green plant materials or a combination of nitrogen poor and nitrogen-rich plant materials does not need mineral N input.

☆ Either spray a suspension of the culture of efficient bio-degrading fungal species characterized as microbial inoculants for rapid composting or mix simply 1 part of fresh cattle dung with 2 parts of the compostable material. Fresh cattle dung serves as highly effective biological inoculant for the composting of organic wastes.

☆ Bring the compost material to about 70 per cent moisture content and carry out composting in pits or heaps, maintaining appropriate moisture level during the composting period.

☆ Add inoculum of *Azotobacter chroococum*, if available, to the composting system after 25-30 days of decomposition.

☆ Provide 4-5 turnings to the composting material at 15 day intervals. A good quality compost rich in humus and plant nutrients will be ready for use in 3-4 months of decomposition.

Appendix-20: Technology for Production of Phospho-compost

☆ Use any organic waste material including crop residues, leaf-litter, partially dried green plant biomass including weeds, animal dung and biodegradable city solid waste. A mixture of various organic materials serves as good composting material as any single material. Rapidity of composting and the quality of the compost improved if fibrous materials like rice straw and sugarcane trash are shredded to 5-6 cm size before being put to composting.

☆ On dry weight basis of the compostable material, fix quantity of Mussoorie rock phosphate either as 12.5 per cent or 25.0 per cent to be incorporated in the compostable organic material. Air-dried organic material may be taken as dry material for calculating the quantity of the rock phosphate to be added for practical purpose. If sufficient quantity of compostable material is available, 12.5 per cent may be chosen as the rock phosphate level. If, however, limited quantity of the organic waste is available, the compostable material should be charged with 25 per cent Mussoorie rock phosphate. Both the levels of rock phosphate will give similar quantity of phospho-compost, only the quantity of phospho-compost to be applied to crops will differ for achieving the same results.

☆ Use the following formula to determine approximate ratios of the various components of phospho-compost:

Organic waste : Animal dung : Soil : Compost/farmyard manure
     8   :   1   : 0.5 :   0.5

☆ These above components serve as composting mix for the production of phospho-compost. Calculate the rock phosphate to be added, on the basis of whole compost mix, consisting of the organic waste, animal dung, soil and mature compost/farmyard manure.

☆ The quantity of the compost mix is reduced to 45 to 55 per cent of its original weight under proper conditions of composting. Thus, if 1,000 kg material is composted, one should expect about 500 kg phospho-compost. This finished compost, on an average, contains about 50 per cent moisture and 50 per cent dry compost. This general rule, gained from research experiments, is useful while considering phospho-compost for field applications.

☆ Make a mixture of the animal dung, soil, farmyard manure and the calculated quantity of rock phosphate, and make a slurry of this mixture in a trough or bucket. Add the slurry to the organic material to be composted and mix the entire mass of the materials as uniformly as possible. Adding rock phosphate to the composting material in the form of slurry ensures very effective transformation of rock phosphate during composting. Maximum transformation of rock phosphate to water-soluble and citric acid soluble forms of P is the hallmark of phosphate.

☆ Make up moisture level of the compost mix to about 70 per cent and mix the material to ensure that it becomes appropriately moist but not excessively wet. Compost the material for 3 months in a pit or heap by maintaining required moisture level. The composting period can be increased if the compostable material is relatively resistant to microbial decomposition. Give 4 turnings to the material with a rake or spade. First turning should be given 10 days after the material is laid for composting, followed by subsequent turnings at 15 day interval. About 75-80 per cent of the achievable decomposition is accomplished in about 60 days of composting. If composting is carried out by maintaining adequate moisture and by providing recommended intermittent turnings, a good phospho-compost, rich in citric acid soluble P, is ready in 3 months, Appropriateness of moisture level should be checked at each turning.

Appendix-21: Biomass in Plantation Crops for Recycling

Plantation Crops	Biomass (t/ha/hr)
Coconut	8-12
Arecanut	8
Locoa	9
Cardamom	5.5
Cashew	1.5
Tea/Coffee	10-20
Shade trees	8-10

Appendix-22: Chemical Composition of Phospho-Compost Compared with Ordinary Compost

Component	Ordinary Compost	Phospho-compost	Rock Phosphate
Total N (per cent)	1.3	0.95	-
pH	8.6	8.8	8.8
Total P (per cent)	0.25	3.00	8.4
Organic P (per cent)	0.05	0.06	0.18
Available P (mg/g)	1.20	0.63	0.18
Water-soluble P (mg/g)	0.31	0.20	-
Citric acid-soluble P (mg/g)	0.52	12.24	2.20

Appendix-23: Chemical Analysis of Organic Fertilizer/Soil Enricher

Chemical Properties	Range
Total organic matter (per cent)	50-65
PH	6.8-7.8
C: N ratio	10: 1-20: 1
N (per cent)	1.80-3.0
P (per cent) total	1.25-2.5
K (per cent)	0.80-1.50
Ca (per cent)	1.5-4.0
Mg (per cent)	0.75 –1.0
S (per cent)	0.40-0.60
Trace elements	All are present
Electrical conductivity	<3.0 dSm^{-1}
Moisture content (per cent) in organic fertilizer pellets	< 10
Moisture content (per cent) in soil enricher (powder)	18-24

Appendix-24: Cation-exchange Capacities of Soil Clays and Humus

Name of Soil/Clay	CEC (Cmol/kg)	
	Average	*Usual Range*
Humus	200	100 – 300
Vermiccullite	150	100 – 200
Allophane	100	50 – 200
Montmorillanite	80	60 – 100
Illite	30	20 – 40
Chlorite	30	20 – 40
Kaolinite	8	3 – 5

Appendix-25: Uptake of Micro-nutrient Cations by some Intensive Cropping Systems under Optimum NPK Rates

Location	Cropping System	Economic Produce (t/ha)	Uptake (g/ha/yr)			
			Zinc (Zn)	Copper (Cu)	Iron (Fe)	Manganese (Mn)
Pusa, Bihar	Rice-Wheat-Jowar	5.98	606	282	4683	1238
Pusa, Bihar	Rice-Mustard-Moong	4.43	445	208	4258	879
Hisar	Rice-Wheat	10.10	250	190	3430	800
Pantnagar	Rice-Wheat-Cow pea fodder	13.73	946	652	5982	1204
Barrackpore	Rice-Wheat-Jute	8.97	787	54	4583	972
Luchiana	Maize-Wheat-Cow pea fodder	9.77	357	98	32.37	870
Delhi	Maize-Wheat-Cow pea fodder	7.32	482	194	3610	820
Jabalpur	Soybean-Wheat	6.07	285	147	1953	615
Hyderabad	Rice-Wheat	7.07	537	1106	2616	2480

Appendix-26: Micro-nutrients Requirements of different Crops

Crops	Iron (Fe)	Manganese (Mn)	Boron (B)	Zinc (Zn)	Copper (Cu)	Molybdenum (Mo)
Cotton	106	14	15	16	8	0.77
Cowpea	260	89	53	17	11	1.31
Greengram (Mung)	170	18	32	13	11	1.05
Sorghum (Jowar)	360	27	27	36	3	0.98
Cabbage	205	28	25	21	8	0.98
Guar	71	15	29	15	9	1.14
Castor	223	41	31	14	2	1.01
Pearlmillet (Bajra)	349	44	-	50	12	-
Hybrid Bajra	264	23	27	22	9	0.84
Groundnut	499	39	44	9	5	1.32
Wheat	232	26	18	21	8	0.87
Potato	160	12	50	9	9	0.80
Brinjal	320	36	45	23	12	0.23
Tobacco leaves	692	132	96	21	11	0.60
Lucerne	420	41	36	21	10	2.59
Maize (Fodder)	217	74	-	30	31	-

Appendix-27: Consumption of Fertilizer (in mt.)

		2003-04		
Zone	*N*	*P_2O_5*	*K_2O*	*Total*
Northern	4493.66	1397.59	229.56	6120.81
Southern	2118.35	921.96	677.87	3718.18
Eastern	1606.92	518.37	361.82	2487.11
Western	2857.41	1285.84	328.30	4471.55
Total	11076.34	4123.76	1597.55	16797.65

Appendix-28: Some Projections on the Availability of Organic Resources for Agriculture in India during 2000-2025

Generators	*2000*	*2010*	*2025*
Human population (million)	1,000	1,120	1,300
Livestock population (million)	498	537	596
Foodgrain production (million t)	230	204	315
Nutrients (theoretical potential)			
Human excreta (million t $N+P_2O_5+K_2O$)	2.00	2.24	2.60
Livestock dung (million t $N+P_2O_5+K_2O$)	6.64	7.00	7.54
Crop residues (million t $N+P_2O_5+K_2O$)	0.21	7.10	20.27
Nutrient (considered tapable)'			
Human excreta (million t $N+P_2O_5+K_2O$)	1.60	1.80	2.10
Livestock dung (million t $N+P_2O_5+K_2O$)	2.00	2.10	2.26
Crop residues (million t $N+P_2O_5+K_2O$)	2.05	2.34	3.39
Total	5.05	6.24	7.75

1. All data pertaining to nutrients in dung and in residues are counted twice to the extent these are fed to the animals.

2. Tapable = 30 per cent of dung, 80 per cent of excreta, 33 per cent of crop residues.

Abbreviations Used in the Text

FYM: Farm yard manure

STCR: Soil test crop response

STR: Soil test based recommendation

RD: Recommended dose

RDF: Recommended dose of fertilizers

GM: Green manuring

OM: Organic manure

INM: Integrated nutrient management

IPNS: Integrated plant nutrient supply system

NPK: Nitrogen phosphorus potassium

CF: Chemical fertilizer

BF: Bio-fertilizers

@: At the rate of

t/ha: Tonnes per hectare

kg: Kilogram

kg/ha: Kilogram per hectare

q/ha: Quintal per hectare

per cent: Percentage of

PSM: Phosphate solubilizing micro-organisms

PSB: Phosphate solubilising bacteria

BGA: Blue green algae

SPS: Species

Mha: Million hectare

P_2O_5: Phosphorus penta Oxide

K_2O: Potassium dioxide

N : P : K: Nitrogen is to (Ratio) Phosphorus is to (Ratio) Potassium

U.P.: Uttar Pradesh

M.P.: Madhya Pradesh

A.P.: Andhra Pradesh

H.P.: Himachal Pradesh

T.N.: Tamil Nadu

J&K: Jammu and Kashmir

SMC: Spent mushroom compost

VAM: Vasicular arbuscular mycorrhizae

SSP: Single super phosphate

DCC: Digested coir pith

FUE: Fertilizer use efficiency

SOM: Soil organic matter

GMO: Genetically modified organism

SOC: Soil organic carbon

LEISA: Low external input sustainable agriculture

Glossary of Soils

A horizon: The surface horizon of a mineral soil having maximum organic matter accumulation, biological activity and/or eluviation of materials such as iron and aluminum oxides and silicate clays.

ABC soil: A soil with a distinctly developed profile, including A,B, and C horizons.

Absorption: The process by which a substance is taken into and included within another substance, *i.e.* intake of water by soil or intake of nutrients by plants.

Absorption, active: Movement of ions and water into the plant root as a result of metabolic processes against an activity gradient.

Absorption, passive: Movement of ions and water into the plant root as a result of diffusion along a gradient.

AC soil: A soil having a profile containing only A and C horizons with no clearly developed B horizon.

Accessory Structural Elements: The nutrients or elements forming parts of more active and vital living plant tissues particularly proteins. For example,

Nitrogen – a constituent of proteins, and constitutes 1-3 per cent of the plant tissue.

Phosphorus – a constituent of nucleoproteins, phytin and lecithin constitutes 0.05 to 1.0 per cent of plant tissues.

Sulphur – a constituent of many proteins; its content varies from 0.05 to 1.50 per cent in plant tissues.

Acid rain: Atmospheric precipitation with pH values less than about 5.6, the acidity being due to inorganic acids such as nitric and sulfuric that are formed when oxides of nitrogen and sulphur are emitted into the atmosphere.

Acid sulphate soils: Soils having sufficient sulphides (FeS_2 and others) and pH < 3 and when drained and aerated enough for cultivation also termed as cat clays.

Acidity active: The activity of hydrogen ion in the aqueous phase of a soil. It is measured and expressed as a pH value.

Acidity, residual: Soil acidity that can be neutralized by lime or other alkaline materials but cannot be replaced by an unbuffered salt solution.

Acidity, salt replaceable: Exchangeable hydrogen and aluminum that can be replaced from an acid soil by an unbuffered salt solution such as KCl or NaCl.

Acid soil: A soil that is acid in reaction throughout the root zone. Practically, this means a soil having pH less than 6.6; precisely, a soil with a pH value less than 7.0. Such a soil has more hydrogen (H) than hydroxyl (OH) ions in the soil solution. Acid soils are grouped into five categories.

Extremely acid: pH below 4.5

Very strongly acid: 4.5-5.0

Strongly acid: 5.1-5.5

Medium acid: 5.6-6.0

Slightly acid: 6.1-6.5.

Acid soils are reclaimed by addition of liming materials.

Acidity, total: The total acidity in a soil. It is approximated by the sum of the salt replaceable acidity plus the residual acidity.

Acid-forming: A term applied to commercial fertilizers which leave an acid residue in the soil. The amount of calcium carbonate required to neutralize the acid residue is referred to its equivalent acidity. Common fertilizers along with their equivalent acidity. Common fertilizers along with their equivalent acidity.

Actinomycetes: A group of microorganisms intermediate between the bacteria and the true fungi that usually produce a characteristic branched mycelium. Includes many, but not all, organisms belonging to the order of Actinomycetales.

Activity index: Index for measuring availability of water insoluble nitrogen in fertilizers. Activity index as per AOAC method is determined by the following formula:

$$\text{Activity index (per cent)} = \frac{\text{(per cent CWIN} - \text{(per cent HWIN)}}{\text{(per cent CWIN)}} \times 100$$

CWIN = cold water insoluble nitrogen

HWIN = Hot water (98 – 100°C) insoluble N

Used as a measure of availability of nitrogen from slow release fertilizers. For urea formaldehyde, which is a slow release fertilizer, activity index should not test less than 40 per cent.

Additive: A material added to fertilizer to improve its physical condition.

Adhesion: It is process of molecular attraction that holds the surfaces of two substances (*e.g.* water and sand particles) in contact.

Adsorption complex: The group of organic and inorganic substances in soil capable of adsorbing ions and molecules.

Adsorption: It is a process of the attraction of ions or compounds to the surface of a solid. Soil colloids adsorb large amounts of ions and water.

Aerate: To impregnate with gas, usually air.

Aeration, soil: The process by which air in the soil is replaced by air from the atmosphere. In a well-aerated soil, the soil air is similar in composition to the atmosphere above the soil. Poorly aerated soils usually contain more carbon dioxide and correspondingly less oxygen than the atmosphere above the soil.

Aerobic: 1) Having molecular oxygen as a part of the environment 2) Growing only in the presence of molecular oxygen, as aerobic organisms 3) Occurring only in the presence of molecular oxygen (said of certain chemical or biochemical processes, such as aerobic decomposition).

Aggregate (soil): Many soil particles held in a single mass or cluster such as a clod, crumb, block, or prism.

Agricultural Chemistry: Includes almost all branches of chemistry, namely, analytical chemistry, inorganic chemistry, organic chemistry, bio-chemistry, radiation chemistry, physical chemistry, as applicable to agriculture system. It is an application of chemistry to soil, water, plant and animal systems. It involves the use of chemistry in fertilizers, amendments, pesticides and insecticides, herbicides and growth hormones.

Agronomy: A specialization of agriculture concerned with the theory and practice of field crop production and soil management. The scientisific management of land.

Air porosity: The proportion of the bulk volume of soil that is filled with air at any given time or under a given condition such as a specified moisture potential, usually the large pores.

Air-dry: 1) The state of dryness (of a soil) at equilibrium with the moisture content in the surrounding atmosphere. The actual moisture content will depend upon the relative humidity and the temperature of the surrounding atmosphere. 2) To allow to reach equilibrium in moisture content with the surrounding atmosphere.

Alkali Soil: A soil that contains sufficient sodium salt to interfere with the growth of most crop plants.

Alkaline forming: A term applied to commercial fertilizers that leaves an alkaline or basic residue in the soil. The basic residual effect is expressed in terms of 'equivalent basicity'. More the basic residue, more will be the equivalent basicity.

Alkaline soil: A soil that is alkaline in reaction throughout the root zone or for a major part of the root zone. Precisely, any soil having a pH value greater than 7.0. Practically, a soil having a pH above 7.3 is called alkaline. An alkaline soil is reclaimed by addition of gypsum or sulphur. Soils having pH from 7 to 7.5 are sometimes referred to as alkali soils. They correspond to black alkali soils and occur in irregular patches. They can be reclaimed by adding calcium or magnesium.

Allophane: An aluminosilicate mineral that has an amorphous or poorly crystalline structure and is commonly found in soils developed from volcanic ash.

Alluvial soil: A soil developing from recently deposited alluvium and exhibiting essentially no horizon development or modification of the recently deposited materials.

Alluvium: A general term for the detrital material deposited or in transit by streams, including gravel, sand, silt, clay and all variations and unconsolidated mixtures of these.

Alpha particle: A positively charged particle (consisting of two protons and two neutrons) that is emitted by certain radioactive compounds.

Aluminosilicates: Compounds containing aluminum, silicon and oxygen as main constituents. An example is microcline, $KAlSi_3O_8$.

AM: Abbreviated terms for 2-ammino-4-chloro-6 methylpyridine used as a nitrification inhibitor. Toxic to bacteria *nitrosomonas* which oxidize ammonium ions to nitrate and hence control nitrification. Rate of application is 4-10 ppm (on soil basis) on mixing with fertilizer. Efficiently reduces nitrification of urea and can also be used as coating material for solid fertilizers using oil as a binder. The effect of AM can last upto 40 days.

Amendment, soil: Any substance other than fertilizers, such as lime, sulfur, gypsum, and sawdust, used to alter the chemical or physical properties of a soil, generally to make it more productive.

Amino acids: Nitrogen-containing organic acids that couple together to form proteins. Each acid molecule contains one or more amino groups ($-NH_2$) and at least one carboxyl group ($-COOH$). In addition, some amino acids contain sulphur.

Ammonification: It is biochemical process whereby ammoniacal nitrogen is released from nitrogen-containing organic compounds.

Ammonium fixation: The entrapment of ammonium ions by the mineral or organic fractions of the soil in forms that are insoluble in water and at least temporarily nonexchangeable.

Ammo-Phos: It is the trade name generally used in the USA and Canada to designate a series of products of commercial monoammonium phosphate, either alone or in combination with ammonium sulphate or other material. Two grades namely Ammo-Phos A and Ammo-Phos B are marketed. Ammo-Phos A is a product of phosphoric acid neutralization with ammonia. It contains 11 per cent N and

48 per cent available P_2O_5. Ammo-Phos B is manufactured by neutralizing phosphoric acid-sulphuric acid mixture with ammonia. It contains 16 per cent N and 20 per cent available P_2O_5.

Ammonium sulphate: Chemically produced sulphur containing nitrogenous fertiliser. A crystalline colourless salt. Due to impurities its colour sometime appears to be brown or grey. Highly soluble in water; not hygroscopic in nature and can be stored and handled without any difficulty. In addition to nitrogen, it also supplies 24 percent sulphur. It is manufactured by three processes. Its specifications as laid down in the Indian Fertiliser Legislation (FCO, 1985) are;

1. Moisture percent by weight maximum: 1.0
2. Ammonical nitrogen percent by weight minimum: 20.6
3. Free acidity (as H_2SO_4) percent by wt., minimum (0.04 for material obtained from by-product ammonia and by product gypsum): 0.025
4. Arsenic (as As_2O_3) percent by weight, maximum: 0.01

All the nitrogen supplied by ammonium sulphate is in ammoniacal form. It is also a carrier of sulphur, hence, favoured for legume crops, tea and coffee. Its equivalent acidity is 110.

Ammonium nitrate: A synthetically produced nitrogenous fertilizer containing one half of its nitrogen in ammonia cal form and the other half in nitrate form. Commercial grade carries 33 per cent nitrogen. It is acid-forming in reaction with 60 equivalent acidity. This fertilizer needs to be stored carefully to avoid fire hazards.

Ammonium phosphate: A series of fertilizer materials containing a chemically combined form of ammonium nitrate and dicalcium phosphate. Two commercial grades of ammonium phosphate being manufactured in India during the Fourth Five Year Plan are 20-20-0 and 28-28-0. These fertilizers are acidic in residual effect. The 20-20-0 grade ammonium phosphate has 25 equivalent acidity.

Ammonium sulphate-nitrate: A synthetically produced nitrogenous fertilizer containing nitrogen in ammoniacal and nitrate form. Commercial grade of this fertilizer contains 26 per cent nitrogen. It is acid-forming in nature, with 93 equivalent acidity.

Anhydrous ammonia: Contains 82 per cent nitrogen in liquid form obtained by compressing ammonia gas. It is highly acidic, with 148 equivalent acidity. It requires special equipment to apply to the soil.

Anion: An ion carrying a negative charge of electricity.

Amphiboles: A group of aluminosilicate minerals having a structure containing double chains of link silica tetrahedral.

Anaerobic: Without molecular oxygen. The opposite of aerobic.

Anion: Negatively charged ion; during electrolysis it is attracted to the positively charged anode.

Anion sorption: The process by which an anion, such as phosphate, replaces OH$^-$ groups on mineral surfaces and edges.

Antagonistic Effect: The application of one nutrient element may accentuate or induce deficiency of another nutrient due to related antagonism. For example, excess copper adversely affects the iron nutrition. Iron and manganese have an antagonistic effect on each other. Zinc and iron behave in a similar way. Heavy doses of phosphorus application induce zinc deficiency. High levels of potassium and calcium depress boron absorption.

AOAC: Abbreviation for Association of Official Analytical Chemists. The new name for the earstwhile American Association which was called the Association of Official Agricultural Chemists. The association was founded in Philadelphia on September 9, 1884. The membership of the association is institutional and includes State Departments of Agriculture, State Agricultural Colleges and Experimental Stations, the Federal Department of Agriculture and the Federal State and City officers charged with enforcement of food, feed, drug, fertilizer, insecticide and fungicide control law.

Ap: The surface layer of a soil disturbed by cultivation or pasturing.

Apatite: A phosphate bearing mineral contained in rock phosphate. There are five types of apatites.

1. Hydroxy apatite $Ca_{10}(PO_4)_6(OH)_2$
2. Chloroapatit $Ca_{10}(PO_4)_6 Cl_2$
3. Fluoroapatite $Ca_{10}(PO_4)_6 F_2$
4. Carbonatoapatite $Ca_{10}(PO_4)_6 CO_3$
5. Sulphatoapatite $Ca_{10}(PO_4)_6 SO_4$

Apatite is a source of phosphorus and phosphoric acid used in fertilizer and other industries. Also used for manufacturing bone chine. Fluorine rich apatite yields hydrofluosilicic acid as a by-product.

AR Grade: An abbreviation for Analytical Reagent Grade. AR grade chemicals are high purity chemical reagents which are used in any standard analytical work requiring high precision. Generally, on packings of such reagents, detailed assay including maximum impurities that can be found in the reagent are given.

Arid climate: Climate in regions that lack sufficient moisture for crop production without irrigation. In cool regions annual precipitation is usually less than 25 cm. It may be as high as 50 cm in tropical regions. Natural vegetation is desert shrubs.

Arnon's criteria of essential elements: Plants absorb a number of elements from soil during their growth. All these elements are not essential for plants. The criteria proposed by Arnon for deciding whether an element is essential for plants or not are;

1. The deficiency of the element makes it impossible for the plant to complete the vegetative or reproductive stages of its life cycle.

2. The deficiency is specific to the element in question: and as such, can be prevented or corrected only by supplying that particular nutrient element to the plant.

3. The element must have a direct influence on the plant and must be directly involved in the nutrition and metabolism of the plant quite apart from its possible effect on correcting some unfavourable microbiological or chemical conditions of the soil or other growing medium.

Atom: The smallest existing unit of an element. The term is derived from the Greak word meaning "individual". It consists of a nucleus which contains protons (positively changed particles) and neutrons (neutral particles). The nucleus is surrounded by electrons (negatively charged particles) which move in their respective orbits. The atom is electrically neutral as the number of electrons in an atom is equal to number of protons.

Atomic Absorption Spectrophotometer (AAS): An instrument used in analysis of a variety of products (plant, soil and water) including fertilizers based on the absorption of light by atoms. More than 60 elements can be determined by AAS.

The basic components of an atomic absorption spectrophotometer are: Atomizer, Light sources (cathode lamps)

Optics, Monochromators and Detector.

Available nutrients in soils: A part of the plant nutrient in the soil that can be taken up by growing plants immediately. Available nitrogen is defined as the water soluble nitrogen plus the part that can be made soluble or converted into free ammonia. Available phosphoric acid is that part which is soluble in water or in a week dilute acid such as 2 per cent citric acid. Available potash is defined as that portion which is soluble in water or in a solution of ammonium oxalate.

Autotrophs: Group of micro organisms which obtain their energy from sunlight or by oxidation of inorganic compounds. If the energy is derived from sunlight, the group is called photo-autotroph. The other group which derives energy from oxidation of inorganic material is called chemo-autotroph. Algae are photo-autotrophs. Some autotroph bacteria which play an important part in availability of plant nutrients are:

1. *Nitrosomonas* – oxidize ammonium to nitrate
2. *Thiobacillus* – oxidize inorganic S to sulphate
3. *Nitrobacters* – oxidize nitrite to nitrate.
4. *T. ferroxidans-oxidise* ferrous iron to ferric form.

Azofication: Non-symbiotic nitrogen fixation by the *azotobacter* group of soil bacteria which use organic matter as a source of energy and are able to obtain nitrogen from the atmosphere to build up their body protein. After the death of the bacterial cell, this nitrogen is returned to the soil for use by higher plants.

Azolla: A small floating aquatic fern that is found in the tropics and sub-tropics and some temperate zones of the world. It is a member of order Salviniales and comprises of seven recognized living species, namely:

1. *Careliniana*
2. *Filiculoides*
3. *Mexicana*
4. *Microphylla*
5. *Nilotica*
6. *Pinnata*
7. *Rubra*

For rice cultivation azolla is applied as green manure both by basal application and as top dressing. Under optimum conditions an azola crop may produce 3 kg or more N/ha/day. Thus a 20 day old azolla crop can produce 60 kg N/ha. Two basal crops of 20 days each thus can produce almost 120 kg N/ha for the following rice crop.

Azolla provides nutrients and protective cavity to anabaena algae, which in exchange fixes atmospheric nitrogen. In addition to nitrogen fixation, the azolla is also used as fodder or feed for animals and as a weed suppressor in flooded rice. In some places like lakes, ponds, waterways and aquatic crops, azolla may be considered as a weed.

Azotobacter: Free living bacteria capable of utilizing atmospheric nitrogen and fixing it for their biosynthetic reaction. These can also utilize ammonium, nitrate, nitrite, urea and sometimes organic nitrogen.

Azotobacter are essentially aerobic. There are five species of azotobacter recognized on the basis of cell shape, pigmentation and mobility:

1. *Azotobacter chroococcum*
2. *Azotobacter beijerinckii*
3. *Azotobacter vinelandii*
4. *Azotobacter macrocytogenes*
5. *Azotobacter agilis*

B horizon: Soil horizon usually beneath the A that is characterized by one or more of the following: q) a concentration of silicate clays, iron and aluminum oxides, and humus, alone or in combination b) a blocky or prismatic structure and c) coatings of iron and aluminum oxides that give darker, stronger, or redder color.

Bacterial culture: Any media enriched with any particular bacteria. The culture describes the bacteria contained in it *e.g.*

1. Rhizobium culture: is used for inoculating seeds of leguminous crops.
2. Azotobacter culture: are of a general nature and same culture can be used for inoculating seeds of a number of crops.

Band placement: The placement of fertilizers in the soil in continuous band or narrow ribbons in between the plant rows or around each plant. The fertilizer applied in band is covered by the soil but is not mixed with it. Phosphatic fertilizers are normally placed in bands to reduce phosphatic fixation.

Bar: A unit of pressure equal to one million dynes per square centimeter (10^6 dynes/cm^2).

Base saturation percentage: The extent to which the adsorption complex of a soil is saturated with exchangeable cations other than hydrogen and aluminum. It is expressed as a percentage of the total cation exchange capacity.

$$\text{Per cent BS} = \frac{BC}{CEC} \times 100$$

Per cent BS = Base saturation

BC = Me of basic cations per 100 gram soil.

CEC=total cation exchange capacity (me/100 g soil)

Basic slag: A by-product of steel industry obtained from phosphate iron ores. It contains about 6 to 18 per cent total phosphoric acid and 5 to 15 per cent in available form (soluble in two per cent citric acid). Basic slag produced in India being of low grade is not used as a commercial fertilizer.

BC soil: A soil profile with B and C horizons but with little or no A horizon. Most BC soils have lost their A horizons by erosion.

Bedrock: The solid rock derlying soils and the regolith in depths ranging from zero (where exposd by erosion) to several hundred feet.

Beer-Lambert law: Beer's law and Lambert's law combined together to form the Beer-Lambert Law. According to this law the magnitude of light absorption is proportional to concentration, thickness of media and intensity of light. Spectrophotometric instruments are based on the Beer-Lambert Law.

Mathematically it is expressed as

$$\text{Log} = \frac{lo}{It}\ KCH$$

lo = Intensity of incident light

It = Intensity of transmitted light

C = Concentration

H = Thickness of media

K = Equilibrium constant depending on the nature of the substance.

Biogas: The gas produced from organic waste by microbiological reaction under anaerobic conditions. The organic wastes used in the production of biogas are generally cattle yard waste, human waste, vegetative crop residues, unwanted aquatic plants and weeds. Biogas is sometimes termed after the material or method of gas production, *e.g.* marsh gas, sewage gas, sludge gas, digestor gas and gobar (cowdung) gas, *etc.* As a fuel, biogas is a potential source of energy. It is usually have the following composing.

Methane = 50-60 per cent

Carbondioxide = 30-40 per cent

Hydrogen = 5-10 per cent

Hydrogensulphide = traces

Water vapour = traces

Biological test: A technique for assessing nutrient status of soil using biological material. The following types of tests are commonly used:

1. Field test: Involves use of field crops in experimental or farmers' fields.

2. Laboratory or green house tests:

    a) Mitscherlich pot culture

    b) Lettuce pot culture

    c) Neubauer seedling method

    d) Sunflower pot culture

3. Microbiological tests:

    a) Azotobacter culture

    b) Sackett and Stewart technique

    c) Aspergillus Niger test

    d) Mehlich's Cunninghamella Plaque method.

Biomass: The amount of living matter in a given area.

Biuret $(C_2O_2N_3H_5)$: A chemical compound formed by the combination of two molecules of urea with the release of a molecule of ammonia, when the temperature during the urea manufacturing process exceeds a certain level. Fertiliser grade urea contains variable amounts of biuret. Biuret is toxic to plants, particularly when it is applied through sprays. As per the Indian fertilizer Legislation (FCO, 1985) the biuret content in urea should not exceed 1.5 per cent.

Blood Meal: Dried blood obtained from a slaughter house. The blood lot is treated with copper sulphate at the rate of 2 oz per 100 lb of blood lot. It is then heated to evaporate from moisture and then dried under the sun. After drying it is powered, bagged and sold as blood meal. Various other treatments like tresting blood serum or clots with lime or very dilute hydrochloric acid or absorbing the blood in organic wastes and then drying in the sun, are also used in the preparation of blood meal. It is quite good organic fertilizer containing 10 to 12 per cent nitrogen and 1-2 per cent P_2O_3.

Blocky soil structure: Soil aggregates with block like shapes, common in B horizons of soils in humid regions.

Blown-out land: Areas from which all or almost all of the soil and soil material has been removed by wind erosion. Usually unfit or crop production. A miscellaneous land type.

Blue-Green Algae: Blue-green algae belong to the class Cyanophyceae or Myxophyceae. These are blue-green in colour because of the presence of pigments like chlorophyll, carotenes, xanthophylls, physcocyanin and phycoerythrin. The majority of blue-green algae species are found in fresh water and are often abundantly found in the standing water of rice fields. They reproduce vegetatively and multiply at a very fast rate. Blue green algae are divided into five orders:

1. *Chroococcales*
2. *Chmaesiphonates*
3. *Plerocapsales*
4. *Nostocales*
5. *Stigonematales*

Important genera concerned with nitrogen-fixing activity are: Anabaena, Nostoc, Cylindrospermum. According to conser-vative estimates, blue-green algae can fix about 25-30 kg N/ha per cropping season.

The algae also give some additional benefits.

1. Because of photosynthetic activity they release oxygen which is a key factor in rice culture.
2. They increase the organic content of soil.
3. They improve the physical structure by increasing water stable aggregates.
4. They supply growth promoting substances like Vitamin B_{12}, IAA-IPA, Anthramilic acid and other allied compounds.

Bonemeal, raw: A fertilizer made of dried animal bones finely ground. It contains 20 to 24 per cent P_2O_5. The availability of its plant food depends largely upon how fine it is ground.

Bonemeal, steamed: A product made from grinding bones previously treated with steam under pressure. It contains one to two per cent nitrogen and 22 per cent phosphorus.

Borax: A chemical compound that contains approximately 11 per cent boron. Borax is applied to boron-deficient soil or sprayed on the plant's foliage which indicate boron deficiency.

Boron: One of the essential elements absorbed by plants as BO_3^{-3} and $HB_4O_2^-$ ions. It is required in a very small quantity, hence is classified as a micronutrient. It is primarily concerned with calcium uptake and its assimilation. It effects cell division, fruit formation, viability of pollen grains, carbohydrate metabolism and protein synthesis in plants. Its deficiency leads to symptoms like heart rot of sugarbeet, browning in cauliflower, corking in apple *etc.*

Bray's nutrient mobility concept: It states that as the mobility of a nutrient in the soil decreases, the amount of that nutrient needed in the soil to produce a maximum yield (the soil nutrient requirement) increases from a variable net value, determined principally by the magnitude of the yield and the optimum percentage composition of the crop, to an optimum percentage composition of the crop, to an amount whose value tends to be constant.

Breccia: A rock composed of coarse angular fragments cemented together.

Buffering capacity: The ability of a soil to resist changes in pH. Commonly determined by presence of clay, humus, and other colloidal materials.

Bulky organic manures: These manures are bulky in nature and supply plant nutrients in small quantities and organic matter in large quantities.

C horizon: A mineral horizon generally beneath the solum that is relatively unaffected by biological activity and pedogenesis and is lacking properties diagnostic of an A or B horizon. It may or may not be like the material from which the A and B have formed.

Calcareous soil: Soil containing sufficient calcium carbonate (often with magnesium carbonate) to effervesce visibly when treated with cold 0.1 N hydrochloric acid.

Calcium ammonium nitrate (CAN): This commercial nitrogenous fertilizer, prepared by mixing powdered limestone or dolomite with ammonium nitrate. It contain 25 percent nitrogen. One half of the nitrogen in calcium ammonium nitrate is in the nitrate form, the remaining half in the ammoniacal form. CAN is suitable for basal dressing as well as for top dressing.

Calcium sulphate or gypsum: Gypsum is hydrated calcium sulphate. The material can be manufactured by treating lime or calcium phosphate with sulphuric acid. It is applied to the soil to reclaim saline and alkaline soils or to bring down soil pH.

Caliche: A layer near the surface, more or less cemented by secondary carbonates of calcium or magnesium precipitated from the soil solution. It may occur as a soft, thin soil horizon, as a hard, thick bed just beneath the solum, or as a surface layer exposed by erosion.

Carbon: An essential plant element. Plants take their carbon requirement from atmospheric carbon dioxide. Since it is never deficient in the atmosphere to become a limiting factor in growth, it is not supplied in any form as fertilizer. Carbon present in organic compounds in soil cannot be used as a nutrient source of carbon.

Carbon cycle: The sequence of transformations whereby carbon dioxide is fixed in living organisms by photosynthesis or by chemosynthesis, liberated by respiration and by the death and decomposition of the fixing organism, used by heterotrophic species, and ultimately returned to its original state.

Carbon-nitrogen ration: The ratio of the weight of total organic carbon to the weight of total nitrogen in a soil or in an organic material. C: N ratio of wheat straw is nearly 80: 1 while that of soil is 10: 1. When undecomposed straw with a high C: N ratio is applied to the soil, its C: N ratio is reduced through bacterial decomposition. To speed up bacterial decomposition, nitrogenous fertilizer is added to the soil at the time of turning under straw.

Catalyst: An element or compound that accelerates the rate of a chemical reaction, but itself does not take part in the reaction. Common characteristics of a catalyst are:

1. The quantity of catalyst required is very small.
2. It does not disturb or has no influence on state of equilibrium of a system.
3. It remains chemically unchanged at the end of a reaction.
4. It cannot start a reaction but can only accelerate the speed of reaction.
5. Catalysts are very specific in nature.

Catalysts are extensively used in the fertilizer industry. The most important example is the use of iron oxide catalyst promoted by metals like aluminium, potassium, calcium and magnesium in the synthesis reaction of ammonia. Nickel is used as a catalyst in steam reforming of natural gas or naptha.

Capillary water: The water held in the "capillary" or *small* pores of a soil, usually with a tension >60 cm of water.

Catena: A sequence of soils of about the same age, derived from similar parent material, and occurring under similar climatic conditions but having different characteristics because of variation in relief and in drainage.

Cation: An ion carrying positive charge of electricity. The common soil cations are calcium, magnesium, sodium, potassium and hydrogen.

Cation exchange: The exchange of cations held by the soil absorbing complex is rich in sodium (as is the case in alkali and alkaline soil), application of gypsum (calcium sulphate) causes calcium cations to exchange with sodium cations.

Cation exchange capacity: The sum total of exchangeable cations that a soil can adsorb. Sometimes called "total-exchange capacity" "base-exchange capacity", or "cation-adsorption capacity." Expressed in centimoles per kilogram (cmol/kg) of soil (or of other adsorbing material such as clay).

Channery: Thin, flat fragments of limestone, sandstone, or schist up to 15 cm (6 in.) in major diameter.

Chelates: An organic compound capable of holding the plant nutrient in a form which prevents it from getting tied with other elements in the soil, thus keeping it more or less in available form for the plant. The term refers to the claws of a crab illustrative of the way in which the atom is held. For examples EDTA, DTPA, HEEDTA and CDTA.

Chert: A structureless form of silica, closely related to flint, that breaks into angular fragments.

Chisel, subsoil: A tillage implement with one or more cultivator-type feet to which are attached strong knife like units used to shatter or loosen hard, compact layers, usually in the subsoil, to depths below normal plow depth.

Chlorite: A 2: 1: 1 type layer structured silicate mineral having 2: 1 layers alternating with a magnesium dominated octahedral sheet.

Chlorosis: A condition in plants relating to the failure of chlorophyll (the green coloring matter) to develop. Chlorotic leaves range from light green through yellow to almost white.

Class, soil: A group of soils having a definite range in a particular property such as acidity, degree of slope, texture, structure, land use capability, degree of erosion, or drainage.

Clastic: Composed of brokwn fragments of rocks and minerals.

Clay: 1) A soil separate consisting of particles <0.002 mm in equivalent diameter 2) A soil textural class containing >40 per cent clay, <45 per cent sand, and <40 per cent silt.

Claypan: A dense, compact, slowly permeable layer in the subsoil having a much higher clay content than the overlying material, from which it is separated by a sharply defined boundary. Claypans are usually hard when dry and plastic and sticky when wet.

Clay humus complex: It is a complex formed due to interaction of the clay and humus colloids through different mechanisms and force like, cation bridge, hydrogen bonding, vander waals forces.

Clod: A compact, coherent mass of soil produced artificially, usually by such human activities as plowing and digging, especially when these operations are performed on soils that are either too wet or too dry for normal tillage operations.

Closed formula mixed fertilizers: The fertilizers grade is disclosed on each bag of such fertilizer mixture, but the ingredients or straight fertilizers used in formulating the mixture are not disclosed. In India, fertilizer mixtures sold to the cultivators are usually of the closed formula type.

Cohension: Holding together: force holding a solid or liquid together, owing to attraction between like molecules. Decreases with rise in temperature.

Colloid, soil (Greek, glue-like) Organis and inorganic matter with very small particle size and a correspondingly large surface area per unit of mass.

Colluvium: A deposit of rock gragments and soil material accumulated at the base of steep slopes as a result of gravitational action.

Complex fertilizers: The commercial fertilizers containing a least two or more of the primary essential nutrients. When such fertilizers contain only two of the primary nutrients, they are designed as incomplete complex fertilizers. While those containing all three primary nutrients are called as complex fertilizers. See multiple-nutrient materials for more information.

Compost: A mass of rotted organic matter made from waste.

Compost: Organic residues, or a mixture of organic residues and soil, that have been piled, moistened, and allowed to undergo biological decomposition. Mineral fertilizers are sometimes added. Often called "artificial manure" or "synthetic manure" if produced primarily from plant residues.

Composting: It is largely a biological process in which micro-organisms of both types namely aerobic and anaerobic, decompose the organic matter and lower the carbon-nitrogen ratio of the refuse. The final product of composting is a well-rooted manure known as compost.

Concretion: A local concentration of a chemical compound, such as calcium carbonate or iron oxide, in the form of grains or nodules of varying size, shape, hardness, and color.

Conditioner (of fertilizer): A material added to a fertilizer to prevent caking and to keep is free-flowing.

Conduction: The transfer of heat by physical contact between two or more objects.

Consumptive use: The net surface charge of mineral particles, the magnitude of which depends only on the chemical and structural composition of the mineral. The charge arises from isomorphous substitution and is not affected by soil pH.

Contour: An imaginary line connecting points of equal elevation on the surface of the soil. A contour terrace is laid out on a sloping soil at right angles to the direction of the slope and nearly level throughout its course.

Convention: The transfer of heat through a gas or solution because of molecular movement.

Copper sulphate: It is the most common copper salt used in fertilizers, insecticides and fungicides. Used as a fertilizer, commercial grade of copper sulphate supplies 25 to 35 per cent copper (Cu).

Cotton seed cake: A by-product of the cotton seed crushing industry, when it is undecorticated, it is non-edible containing 3.9 per cent nitrogen, 18 1.8 per cent phosphoric acid and 1.6 per cent potash. When decorticated, cotton seed cake becomes edible oilcake containing higher percentages of plant nutrients. Decorticated cotton seed cake contains 6.4 per cent nitrogen, 2.9 per cent phosphoric acid and 2.2 per cent potash.

Crop rotation: A planned sequence of crops growing in a regularly recurring succession on the same area of land, as contrasted to continuous culture of one crop or growing different crops in haphazard order.

Crust: A surface layer on soils, ranging in thidkness from a few millimeters to perhaps as much as 3 cm, that is much more compact, hard, and brittle, when dry, than the material immediately beneath it.

Darcy's law: Darcy stated that the rate of flow increased with an increased depth of water above the bottom of the soil and decreased with an increased depth of soil through which water followed and

$$Qw = \frac{-K \,(\Delta dw)\, At}{\Delta ds}$$

Qw = Quantity of water in c.c.

K = Hydraulic conductivity cm/sec.

Δdw = Water of hydraulic head in cm.

A = Soil area in sq. cm.

t = time in second

Δds = soil depth in cm.

Deficiency: A condition of insufficient supply of essential plant nutrients required for various metabolic functions.

Plants cannot reach optimum growth, if any one of the essential elements is limited in supply. Deficiency may be caused due to any one of the following reasons:

1. The nutrient is in short supply in soil.
2. The nutrient is there in soil, but is not in the available form in which plants can absorb it.
3. There is the antagonistic effect of other elements. See also Antagonistic effect. This is called induced deficiency as the nutrient in question is not being metabolized because of the excess of other antagonistic elements.

Denitrification: The process by which nitrates or nitrites in the soil or in farmyard manure are reduced to ammonia or free nitrogen by bacterial action. The process results in the escape of nitrogen into the air and is therefore wasteful.

Desalinization: Removal of salts from saline soil, usually be leaching.

Diagnostic subsurface horizons: The following diagnostic subsurface horizons are used in *Soil taxonomy*.

☆ *Agric horizon:* A mineral soil horizon in which clay, silt, and humus derived from an overlying cultivated and fertilized layer have accumulated. The wormholes and illuvial clay, silt, and humus occupy at least 5 per cent of the horizon by volume.

☆ *Argillic horizon*: A mineral soil horizon characterized by the illuvial accumulation of layer lattice silicate clays.

☆ *Calcic horizon:* A mineral soil horizon of secondary carbonate enrichment that is more than 15 cm thick, has a calcium carbonate equivalent of more than 15 per cent and has a least 5 per cent more calcium carbonate equivalent than the underlying C horizon.

☆ *Cambic horizon*: A mineral soil horizon that has a texture of loamy very fine sand or diner, contains some weatherable minerals, and is characterized by the alteration or removal of mineral material. The cambic horizon lacks cementation or induration and has too few evidences of illuviation to meet the requirements of the argillic or spodic horizon.

☆ *Duripan:* A mineral soil horizon that is cemented by silica, to the point that air-dry fragments will not slake in water or HCl.

☆ *Gypsic horizon*: A mineral soil horizon of seconday calcium sulfate enrichment that is more than 15 cm thick.

☆ *Kandic horizon:* A horizon having a sharp clay increase relative to overlying horizons and having low-activity clays.

☆ *Natric horizon*: A mineral soil horizon that satisfies the requirements of an argillic horizon, but that also has prismatic, columnar, or blocky structure and a subhorizon having more than 15 per cent saturation with exchangeable sodium.

☆ *Oxic horizon*: A mineral soil horizon that is at least 30 cm thick and characterized by the virtual absence of weatherable primary minerals or 2: 1 latice clays and the presence of 1: 1 lattice clays and highly insoluble minerals such as quartz sand, hydrated oxides of iron and aluminum, low cation exchange capacity, and small amounts of exchangeable bases.

☆ *Spodic horizon:* A mineral soil horizon characterized by the illuvial accumulation of amorphous materials composed of aluminum and organic carbon with or without iron.

Diatomaceous earch: A geologic deposit of fine, grayish, siliceous material composed chiefly or wholly of the remains of diatoms. It may occur as a powder or as a porous, rigid material.

Diatoms: Algae having siliceous cell walls that persist as a skeleton after death, any of the microscopic unicellular or colonial algae constituting the class Bacillariaceae. They occur abundantly in fresh and salt waters and their remains are widely distributed in soils.

Diffusion: The transport of matter as a result of the movement of the constituent particles. The intermingling of two gases or liquids in contact with each other takes place by diffusion.

Dilution: The process of lowering the concentration of any component of mixture (say a solution, slurry, suspension or solid) by adding any other suitable substance called a diluant. It is very common practice undertaken in any chemical laboratory and is done to bring the elemental concentration within the working range of the equipment or method. For instance, zinc sulphate heptahydrate is guaranteed to contain 21 per cent or 210000 ppm zinc. To bring it to the working range (about 1 ppm Zn) of the atomic absorption spectrophotometery it is diluted 2,00,000 times by adding distilled water.

Dioctahedral: An octahedral sheet, or a mineral containing such a sheet, that has two thirds of the octahedral sites filled with trivalent ions such as aluminum or ferric iron.

Disintegration: The breakdown of rock and mineral particles into smaller particles by physical forces such as frost action.

Disperse (1) To break up compound particles, such as aggregates, into the individual component particles. (2) To distribute or suspend fine particles, such as clay, in or throughout a dispersion medium, such as water.

Diversion dam: A structure or barrier built to divert part or all of the water of a stream to a different course.

Dolomite: A natural mineral composed of calcium and magnesium carbonate widely used as liming material and as an ingredient in fertilizer mixtures. Dolomite is used in fertilizers to render them non-acid forming. It also supplies available magnesium.

Drain (1) To provide channels, such as open ditches or drain tile, so that excess water can be removed by surface or by internal flow (2) To lose water (from the soil) by percolation.

Drainage, soil: The frequency and duration of periods when the soil is free from saturation with water.

Drift: Material of any sort deposited by geological processes in one place after having been removed from another. Glacial drift includes material moved by the glaciers and by the streams and lakes associated with them.

Drumlin: Long, smooth cigar-shaped low hills of glacial till, with their long axes parallel to the direction of ice movement.

Dry farming: The system in which field crops are raised with an annual rainfall of less than 25 inches, without irrigation facilities. From plant nutrition point of view, lesser doses of fertilizers are recommended to dry farming areas for all field crops, compared to irrigated farming.

Dryland farming: The practice of crop production in lo rainfall areas without irrigation.

DTPA: It is abbreviated term of diethylene trinitrilopenta acetic acid. It is a chelating agent and used for making metal chelates keeping the different metellic cations available in the soil and also increases the translocation of metallic cations within the plant body.

Duff: The matted, partly decomposed organic surface layer of forest soils.

Duripan: A mineral soil horizon that is cemented by silica, to the point that air-dry fragments will not slake in water or HCl.

Edaphology: The science which deals with the influence of soil on living things, particularly plants. It also include the use of soil by men for plant growth.

Efficacy of fertilizer: The whole of the amount of nutrient from the applied fertilizer is not recovered by the crop as a part of it is lost or remains in the soil. The extent of recovery of applied nutrient by a crop or crop rotation indicates the efficacy of a fertilizer. Some factors for controlling efficiency of fertilizer application are listed below:

    ☆ The nature of crop and its variety.

    ☆ Method and time of application of fertilizer.

    ☆ Crop management

    ☆ Cropping system.

    ☆ Chemical composition of soil and its pH.

    ☆ Organic matter content of the soil.

    ☆ Physical condition including drainage, aeration, *etc.*

    ☆ Weather conditions.

    ☆ Soil moisture.

    ☆ Balance of nutrients.

E horizon: Horizon characterized by maximum illuviation (washing out) of silicate clays and iron and aluminum oxides, commonly occurs above the B horizon and below the A horizon.

Earth flow: The process of saturated soil moving down a slope under the force of gravity. The term is also used to describe the results of the process.

Effective precipitation: The amount of proportion of precipitation that infiltrates into soil.

Effluent: Liquid waste from either a septic tank or a sewage treatment plant.

Eutrofication: Pollution with unwanted nutrients.

Electromagnetic spectrum: Range of energy and wavelengths of electromagnetic radiation. The range from shortwave high energy radiation to longwave low energy radiation includes X-ray, ultraviolet, visible light, infrared and radio waves.

Electron: A small, negatively charged atomic particle.

Electrostatic (Attraction or repulsion): Interaction between electrically charged objects.

Element: Any substance that cannot be further separated except by nuclear disintegration.

Eluvial horizon: A soil layer (horizon) formed by the removal of constituents such as clay or iron.

Eluviation: Process of removing from a soil layer any soil constituents in suspension.

Emissivity: Relative measure of an objects ability to emit radiant energy at a given temperature.

Emulsion: The light-sensitive coating on photographic film, usually containing a silver salt such as silver chloride.

Enzymes: Protein catalysts produced in cells of living oranisms that direct and control the cells chemical reactions.

Eolian soil material: Soil material accumulated through wind erosion.

Epidermis: In plants, the outside layer of cells.

Epipedons: Surface layers of soil with specific characteristics used in classifying soils by soil taxonomy (*e.g.,* the mollic epipedon).

Equivalent: The unit formerly used to describe cation exchange capacity or quantities of ions. One gram atomic weight of hydrogen or the amount of any other ion that will combine with or displace this amount of hydrogen. The amount that provides 1 mol of charge. Equals 1 mol.

Equivalent acidity: A term used to express the calcium carbonate equivalent of acidic residue left by a fertilizer material.

Equivalent basicity: A term used to express the calcium carbonate equivalent of basic residue left by a fertilizer material See 'alkaline-forming' for further details.

Equivalent weight: The weight of an element that will combine with 7.9997 of oxygen or 1.0079 of hydrogen. Equivalent weight of molecules is calculated by dividing molecular weight with the number of hydrogen cations or protons/ electrons that will take part in any reaction with the molecule, *e.g.* Equivalent weight of sulphuric acid = 98/2 = 49 where 98 is the molecular weight of sulphuric acid and 2 is the number of hydrogen cations that will take part in any reaction with sulphuric acid.

Erosion: The wearing away of the land surface by water, wind, ice, or gravity.

Essential plant nutrient: A nutrient essential for plants for proper growth and development. At present, there are sixteen essential plant nutrients recognized by plant physiologists. These are carbon, hydrogen, oxygen, nitrogen, phosphorus, calcium, potassium, magnesium, sulphur, manganese, boron, copper, zinc, iron, molybdenum and chlorine.

Eucaryotic: In biology, referring to the type of cells with a distinct nucleus with a nuclear membrane, characteristic of fungi, protozoa, algae, plants and animals.

Evapotranspiration: Evaporation plus transpiration.

Exchange complex: All the materials (clay, humus) that contribute to a soils exchange capacity.

Exchangeable base saturation: Exchangeable cations other than Al^{3+} and H^+, expressed as a percentage of cation exchange capacity measured at neutrality.

Exchangeable ions: Ions (charged atoms or molecules) held by electrical attraction at charged surfaces, can be displaced by exchange with other ions.

Exchangeable potassium: That form of potassium is soil exchange complex which is not recovered by water solution, but is recovered by the exchange process using a suitable reagent. The most commonly used reagent is ammonium acetate. Exchangeable potassium is the primary source of potassium for plants. The level of exchangeable potassium in soil is an important factor in determining the responsiveness of any crop to application of potassic fertilizers.

Exchangeable sodium percentage (ESP): Amount of exchangeable Na expressed as a percentage of total exchangeable cations. It can be expressed by given formula.

$$ESP = \frac{\text{Exchangeable sodium (Cmol/kg soil)}}{\text{Cation exchange capacity (Cmol/kg soil)}} \times 100$$

Extracellular: Outside the cell. Extracellular enzymes are excreted by some bacteria and fungi.

Extracellular enzyme: A protein substance that acts as an organic catalyst excreted outside the bacterial or plant cell. Also called an *exoenzyme.*

Extrusive rock: Any rock that forms from hot molten material by solidifying at the earths surface

Facultative organism: An organism capable of both aerobic and anaerobic metabolism.

Fallow: Cropland left idle in order to restore productivity, mainly through accumulation of water, nutrients, or both. Summer fallow is a common stage before cereal grain in regions of limited rainfall. The soil is kept free of weeds and other vegetation, thereby conserving nutrients and water for the next years crop.

Fallowing: Keeping the land free of a crop or weeds for a period of time. This is done to restore soil productivity mainly through accumulation of water, nutrients or both. Fallow during the kharif season is a common practice in wheat-growing regions which receive limited rainfall. During fallow, the field is cultivated to control weeds and to help the storage of moisture for the succeeding crop.

Family, soil: In soil classification, one of the categories intermediate between the great group and the soil series. Families are defined largely on the basis of physical and mineralogical properties of importance to plant growth.

Farmyard manure (FYM): It is a product of decomposition of the liquid and solid excreta of animals stored in the farm. In western countries, straw or other litter used as bedding is also included along with the animal excreta. In India, since straw is used mainly for fodder purposes, farmyard manure is made mainly from animal excreta. The composition of FYM varies with the nutrient content in excreta and with the method of preparation. On an average it contains 0.5 per cent N, 0.2 per cent P_2O_5 and 0.5 per cent K_2O.

Fauna: The animal life of a region.

Fermentation: A set of metabolic processes by which anaerobic organisms obtain energy by converting sugars to alsohols or acids and CO_2.

Ferrihydrite: A dark raddish brown poorly crystalline iron oxide that forms in wet soils.

Fertigation: The addition of soluble fertilizer through an irrigation system.

Fertility, soil: The quality of a soil that enables it to provide essential chemical elements in quantities and proportions for the growth of specified plants.

Fertility index: It is defined as the relative sufficiency expressed as a percentage of the amount of nutrient adequate for optimum yields. It is related with soil test values and crop response as follows:

Soil test rating	Crop response	Fertility index
Very low	Highly probable	0-10
Low	Probable	10-25
Medium	Possible	25-50
High	Unlikely	50-100
Very high	Highly unlikely	100+

Fertilizer (Control) order, (1985): An order issued by the Government of India under powers conferred by section-3 of the Essential Commodities Act, 1955. Under the powers, the Government controls the production, marketing, price and quality of fertilizers. There are 39 clauses of this order covering definitions, prices, registration of dealers and fertilizer mixtures, regulation on manufacture, sale, packing requirement, disposal of non-standard fertilizers, enforcement authorities, analysis of samples, *etc*. The Schedule I of the order includes detailed specifications of fertilizer covered by it. In Schedule II detailed procedure regarding sampling technique and methods of analysis is given.

Fertilizer grade: An expression that indicates the percentage of plant nutrients in a fertilizer, *e.g.* a 10-55 grade of fertilizer indicates 10 per cent nitrogen, 5 per cent phosphoric acid (P_2O_5) and 5 per cent potash (K_2O).

Fertilizer legislation: This refers to laws and regulations enforced by the Government to regulate the quality of fertilizers or fertilizer mixtures sold to consumers. According to the Fertilizer Control Order of the Govt.of India of 1957, each brand and grade of commercial fertilizer and fertilizer mixture is to be registered and each bag of fertilizer to be labeled with fertilizer grade clearly written on it.

Fertilizer ratio: The fertilizer ratio designates the relative proportion of three major plant nutrients, keeping the percentage of nitrogen as one of the ratio. Thus a 5-10-5 fertilizer mixture which contains 5 per cent total nitrogen. 10 per cent available P_2O_5 and 5 per cent soluble potash, has a nutrient ratio 1: 2: 1.

Fertilizer requirement: The quantity of certain plant nutrient elements needed, in addition to the amount supplied by the soil, to increase plant growth to a designated optimum.

Fertilizer: Any natural or manufactured material, dry or liquid, added to the soil in order to supply one or more plant nutrients. The term is generally applied to commercially manufactured materials other than lime or gypsum. When fertilizers are sold on a large scale, they are called commercial fertilizers. Fertilizers consumed in India on a large scale are of five types; nitrogenous fertilizers, phosphatic fertilizers, potassic fertilizers, complex fertilizers or fertilizers supplying more than one major plant nutrient and fertilizer mixtures.

Field capacity (field moisture capacity): The percentage of water remaining in a soil two or three days after its harving been saturated and after free drainage has practically ceased.

Field experiments: Experiments conducted in the field to determine the type and amount of fertilizers to suit particular soil type and crop.

Filler: Material added to a fertilizer mixture or fertilizer material to make up the difference between the weight of an added ingredient required and the total fertilizer that should be made according to the fertilizer formula. Materials commonly used as fillers are sand, dolomitic limestone, gypsum, furfural, vermiculite, rock salt, coal ashes, chalk grit, spent foundry sand, peanut hulls, saw dust and rice husk, *etc*.

Fixation (in soil): Conversion of a soluble material such as a plant nutrient like phosphorus, from a soluble or exchangeable form to a relatively insoluble form. To reduce fixation of phosphate, phosphatic fertilizers are brought in lesser contact with the soil particles and applied closer to the plant root through band placement.

Flocculation: Joining of colloidal particles to form clusters (flocs).

Foliar diagnosis: Diagnostic technique based on analyzing plant tissues for the total content of nitrogen, phosphorus, potassium, boron, *etc.* Low values of nutrient percentages in plants under test compared to nutrient percentage in high yielding plants indicate deficiency of nutrients. The nutrients commonly tested by this method are nitrogen, phosphorus and potash.

Foliar fertilization: Fertilization of plants, or feeding nutrients to plants, by applying chemical fertilizers to be foliage. This is also known as foliar feeding or spray fertilization.

Fortification: The process of putting an additional quantity of fertilizer to increase the nutrient content of manure/compost, *e.g.* in the ADCO process of compost-making superphosphate is added to fortify the phosphate content of the manure.

Frictional resistance: Resistance to the movement of fluids or particles caused by the interaction (rubbing) of surfaces.

Gamma ray: A high energy ray (photon) emitted during radioactive decay of certain elements.

Geostatistics: Statistics that describe the variability of properties from place to place.

Geographical Information system (GIS): It is a computerised data base management system for capturing, storing, validation, analysing, displaying and managing spatially referenced data sources in addition to the primary data such as agro-climatic and soil characteristics. It can be used to generate map on erosion hazard, land suitability for a specified alternative land use type.

Genesis, Soil: The mode of origin of the soil, with special reference to the processes responsible for the development of the solum, or true soil, from the unconsolidated parent material.

Geological erosion: As opposed to accelerated erosion, erosion at natural rates, unaffected by human activity.

Geothite (FeOOH): A yellow-brown iron oxide mineral that account for the brown colour in many soils.

Gibbsite Al(OH)$_3$: An alluminium trihydroxide mineral most common in highly weathered soils such as oxisols.

Glacial outwash: Geological material moved by glaciers and subsequently sorted and depositied by streams flowing from melting ice. Also called *glaciofluvial deposits.*

Glacial till: Unsorted and unstratified geological material deposited directly by glacial ice.

Glaciation: The process of geological erosion by means of glacial ice.

Gleying: A process that produces reduction of iron and other elements under conditions of prolonged saturation.

Granular fertilizer: A fertilizer composed of particles of roughly the same composition, and about one-tenth of an inch in diameter. This kind of fertilizer is superior in efficiency compared to the fine or powdery fertilizer due to ease in handling and less fixation in soil.

Green leaf manuring: This refers to turning under of green leaves and tender green twigs collected from shrubs and trees grown on bunds, waste lands and nearby forest areas. The common shrubs and trees useful for this purpose are glyricidia *(Glyricidia maculate)*, *Sesbania speciosa*, karanj *(Pongamia pinnata)* etc.

Green manure crop: Any crop grown and buried into the soil for improving the soil condition by addition of organic matter. Such crops are legumes and non-legumes, but mostly legumes *e.g.* sannhamp *(Crotolaria juncia)*, Dhaincha – *(Sesbania aculeata)*.

Green manuring: A practice of ploughing or turning into the soil undecomposed green plant material for improving the physical condition of the soil or for adding nitrogen when the green manure crop is legume. Two types of green manuring are being practiced by the cultivators of India. These are i) green manuring in situ and ii) Greenleaf manuring.

Green manuring in situ: A practice in which green manure crops are grown and buried in the same field which is to be green-manured. The crops are grown alone or intercropped with the main crop. Important green manure crops used in this system are sannhemp *(Crotalaria juncea)*, dhaincha *(Seshania aculeate)*, pillipesara *(Phaseolus trilobus)*, and guar *(Cyamopsis psoraloides* or *tetragonoloba)*.

Greenhouse effect: The warming of the earth's surface and atmosphere owing to absorption of out-going radiation by CO_2, CH_4 and H_2O (like absorption by glass).

Groundnut cake: A by-product of the oil industry. This is an edible oil cake. It contains 7.3 per cent nitrogen, 1.5 per cent phosphoric acid and 1.3 per cent potash. In the Deccan tract of Maharashtra, Karnataka, Andhra Pradesh and Tamil Nadu, groundnut cake is applied to sugarcane and bananas on a very large scale. Besides the nutrients, it supplies organic matter to the soil.

Guano: In some parts of America used synonymously with fertilizer. It is derived from the Spanish word meaning dung. Guano is made up of excrement of seafowl, together with their body remains. Deposits of guano found in many islands may vary from a few inches to over 200 feet in thickness. Peru is one of the pioneer producers and exporters of guano. It is also obtained from excrement and dead remains of sea creatures other than seafowl. It is also called bat guano, seal guano, fish guano, whale guano, *etc.* Some guanos produced from sheep and goat herds are called sheep guano or goat guano. The colour of guano may vary from grey to dark brown. The chemical composition of guano also varies. Its nitrogen content may vary from 4 to 16 per cent and total P_2O_5 may range between 12 and 26 per cent. When nitrogen content is very high, say, 8-16 per cent, it is termed as nitrogenous guano. When phosphorus content is very high, say, 20-25 per cent P_2O_5, it is termed as phosphatic guano. The potash (K_2O) content of guano is 2 to 3 per cent.

The product obtained after treatment of guano with sulphuric acid is called dissolved guano. The available phosphorus and nitrogen content increase as a result of this treatment. Some times any one or all of the three nutrients, namely, N, P_2O_5 and K_2O are added to guano to make it a balanced fertilizer. This is called rectified or fortified guano.

Gypsum: Hydrated calcium sulphate ($CaSO_4$, $2H_2O$) also known as land plaster. Agricultural grade gypsum contains 55 to 95 per cent gypsum. Gypsum is mined from sedimentary deposits. It is also obtained as a by product of phosphoric acid industry and is termed as phospo-gypsum. Gypsum is calcimined to produce plaster of paris which has wide industrial usage. It is also used in cement, pain and many other industries. In agriculture its main use is in reclamation of alkali soils. It is also a source of calcium, which is an essential plant element.

Halophyte: A plant that requires or tolerates a saline (high salt) environment.

Hardpan: A hardened soil layer, in the lower A or in the B horizon, caused by cementation of soil particles with organic matter or with materials such as silica, sesquioxides, or calcium carbonate. The hardness does not change appreciably with changes in moisture content and pieces of the hard layer do not slake in water.

Harrowing: A secondary broadcast tillage operation that pulverizes, smooths, and firms the soil in seedbed preparation, controls weeds, or incorporates material spread on the surface.

Heaving: The partial lifting of plants, buildings, road-ways, fenceposts, *etc.*, out of the ground, as a result of freezing and thawing of the surface soil during the winter.

Heavy metals: Metals with particle densities > 5.0 mg/m^3.

Heavy soil: A soil with a high content of the fine separates, particularly clay, or one with a high drawbar pull, hence difficult to cultivate.

Hematite, Fe_2O_3: A red iron oxide mineral that contributes red color to many soils.

Hemicellulose: A group of complex carbohydrates; polysaccharides that, unlike starch and cellulose, contain other sugars besides glucose. Important to plant cell walls.

Herbicide: A chemical that kills plants or inhibits their growth; intended for weed control.

Herbivore: A plant-eating animal.

Heterotroph: An organism capable of deriving energy for life processes only from the decomposition of organic compounds and incapable of using inorganic compounds as sole sources of energy or for organic synthesis.

High-analysis fertilizer material: The fertilizers containing a higher percentage of the plant nutrients. Such as urea containing 46 per cent nitrogen, triple superphosphate containing 45 to 47 per cent P_2O_5 and muriate of potash containing 60 per cent K_2O. High-analysis fertilizers have an advantage in the cost of bagging, handling and transportation per unit of plant.

Hill placement: This refers to applying fertilizer either in bands or near plants. This method is usually employed when relatively small quantities are to be applied. In Egypt, nitrogenous fertilizer is applied in pinches at the base of each plant of cotton. Maize, cauliflower, cabbage, brinjal, tomato, and other vegetable crops are fertilized by this method.

Horizon, soil: A layer of soil, approximately parallel to the soil surface, differing in properties and characteristics from adjacent layers below or above it.

Horn-and-hoof-meal: A manure prepared from horns and hoofs of animals. Such a meal contains about 13 per cent nitrogen.

Horticulture: The art and science of growing fruits, vegetables, and ornamental plants.

Humic acid: A mixture of variable or indefinite composition of dark organic substances, precipitated upon acidification of a dilute alkali extract from soil.

Humid climate: Climate in regions where moisture, when distributed normally throughout the year, should not limit crop production. In cool climate annual precipitation may be as little as 25 cm; in hot climates, 150 cm or even more. Natural vegetation in uncultivated areas is forests.

Humification: The process involved in the decomposition of organic matter and leading to the formation of humus.

Humin: The fraction of the soil organic matter that is not dissolved upon extraction of the soil with dilute alkali.

Humus: That more or less stable fraction of the soil organic matter remaining after the major portions of added plant and animal residues have decomposed. Usually it is dark in colour. The term is often used synonymously with soil organic matter.

Hydrated: Having water attached or incorporated as part of a chemical substance.

Hydration: The chemical combination of water with another substance.

Hydraulic conductivity: An expression of the readiness with which a liquid such as water flows through a solid such as soil in response to a given potential gradient.

Hydrologic cycle: The circuit of water movement from the atmosphere to the Earth and back to the atmosphere through various stages or processes, as precipitation, interception, runoff, infiltration, percolation, storage, evaporation, and transpiration.

Hydrolysis: A reaction that involves products (H^+ or OH^-) of the dissociation of water. A mineral weathering reaction that adds H^+ to a mineral structure.

Hydromulching: Technique of spraying a slurry of fiber, seed, fertilizer, and chemicals onto roadsides for erosion control.

Hydroscopic coefficient: The amount of moisture in a dry soil when it is in equilibrium with some standard relative humidity near a saturated atmosphere (about 98 per cent), expressed in terms of percentage on the basis of oven-dry soil.

Hydrous oxides, hydroxyoxides: Sesquioxides. Oxides of Fe, Al, or similar metals, with different proportions of water and hydroxyl in the structure.

Hydroxyl: A OH ion or group.

Hygroscopic: A term used to designate materials that absorb moisture from the air.

Hygroscopicity of fertilizer materials: Some fertilizer materials absorb moisture from the air, which causes them to become sticky. Such hygroscopic fertilizers are calcium ammonium nitrate, sodium nitrate, ammonium chloride and urea.

Igneous Rock: The earliest rocks formed by cooling of the earth's molten magma. Based on silicic acid content igneous rocks are classified in four groups:

1. Acid rocks: Silicic acid content is more than 65 per cent, e.g granite.
2. Sub-Acid Rocks: Silicic acid content is 60-65 per cent, *e.g.* syenite.
3. Sub-Basic Rocks: Silicic acid content is 55-60 per cent, e.g diorite.
4. Basic Rocks: Silicic acid content is 40-45 per cent, e.g golomite, basalt.

Illuvial horizon: A soil layer or horizon in which material carried from an overlying layer has been precipitated from solution or deposited from suspension. The layer of accumulation.

Immature soil: A soil with indistinct or only slightly developed horizons because of the relatively short time it has been subjected to the various soil-forming processes. A soil that has not reached equilibrium with its environment.

Immobilization: The conversion of an element from the inorganic to the organic form in microbial tissues or in plant tissues, thus rendering the element not readily available to other organisms or to plants.

Imogolite: A poorly crystalline aluminosilicate mineral with an approximate formula $SiO_2Al_2O_3.5H_2O$; occurs mostly in soils formed from volcanic ash.

Impermeable: Unable to transmit water. A term of ten used with dense soil horizons through which water moves extremely slowly.

Impervious: Resistant to penetration by fluids or by roots.

Inactivated organisms: In waste disposal, harmful organisms made harmless by reaction with the soil and soil organisms.

Inclusion: One or more polypedons within a map unit, not identified by the map unit name.

Indicator plant: A plant that indicates, in general, and often in a specific manner, a deficiency of plant nutrient in the soil.

Indurated (soil): Soil material cemented into a hard mass that will not soften on wetting.

Inferred properties: Soil properties that are not seen or measured but are assumed from other properties. Suitability for a use is an inferred property.

Infiltration rate: A soil characteristic determining or describing the maximum rate at which water can enter the soil under specified conditions, including the presence of an excess of water.

Infrared: Refers to the electromagnetic radiation of wavelength longer than light but shorter than radio. Sometimes called heat radiation.

Inhibitor: The opposite of catalyst as it retards or reduces the rate of a chemical reaction. It is also known as a retarder. Antioxidants are well known inhibitors of oxidation reactions. Benzoic acid and salicylic acid are added to food products to prevent rancidity.

Inoculation: The process of introducing pure or mixed cultures of microorganisms into natural or artificial culture media.

Inorganic compounds: All chemical compounds in nature except compounds of carbon other than carbon monoxide, carbon dioxide, and carbonates.

In-place parent material: Parent material that has not been transported from its original location, for example, a bedrock is an in-place parent material.

Insecticide: A chemical that kills insects.

Insulator: A material (or space) that transmits heat or electricity poorly.

Interaction: A combined effect of two or more factors or variables such that the variables modify each other's effects.

Interaction, electrostatic: Adsorption caused by the electrical attraction of ions to a charged surface.

Intercrop: Two or more crops grown together on the same piece of land at the same time.

Intergrade: A soil that possesses moderately well developed distinguishing characteristics of two or more genetically related great soil groups.

Integrated plant nutrition system (IPNS): It is a concept proposed by FAO where the basic goal is the maintenance or adjustment and possibly improvement of soil fertility and of plant nutrient supply to an optimum level for sustaining the desired crop productivity through optimization of the benefits from all possible sources of plant nutrients in an integrated manner.

Interlayer (mineralogy): Materials between layers within a given crystal, including cations, hydrated cations, organic molecules, and hydroxide groups or sheets.

Interveinal: Between veins.

Interveinal chlorosis: Chlorosis only between leaf veins.

Intrusive rock: Any rock that forms by cooling from a hot molten mass below the earth's surface. Granite is an intrusive rock.

Ion exchange: The interchange between an ion in solution and another ion on the surface of any surface-active material such as clay or humus.

Ions: Electrically charged particles. As used in soils, an ion refers to an electrically charged element, resulting from the breaking up of many substances like acids, salts, and alkalines. Positively charged ions are called 'cations' and negatively charged as 'anions'. For further information see cation and anion.

Iron-pan: An indurated soil horizon in which iron oxide is the principal camenting agent.

Irrigation efficiency: The ratio of the water actually consumed by crops on an irrigated area to the amount of water diverted from the source onto the area.

Irrigation methods: Methods by which water is artificially applied to an area. The methods and the manner of applying the water are as follows.

    ☆ *Border-strip:* The water is applied at the upper end of a strip with earth borders to confine the water to the strip.

    ☆ *Center pivot*: Automated sprinkler irrigation achieved by automatically rotating the sprinkler pipe or boom, supplying water to the sprinkler heads or nozzles, as a radius from the center of the field to be irrigated.

    ☆ *Check-basin:* The water is applied rapidly to relatively level plots surrounded by levees. The basin is a small check.

    ☆ *Corrugation:* The water is applied to small, closely spaced furrows, frequently in grain and forage crops, to confine the flow of irrigation water to one direction.

    ☆ *Drip*: A planned irrigation system where all necessary facilities have been installed for the efficient application of water directly to the root zone of plants by means of applicators (orifices, emitters, porous tubing, perforated pipe, *etc.*) operated under low pressure. The applicators may be placed on or below the surface of the ground.

    ☆ *Flooding*: The water is released from field ditches and allowed to flood over the land.

    ☆ *Furrow:* The water is applied to row crops in ditches made by tillage implements.

    ☆ *Spinkler:* The water issprayed over the soil surface through nozzles from a pressure system.

    ☆ *Subirrigation*: The water is applied in open ditches or tile lines until the water table is raised sufficiently to wet the soil.

    ☆ *Wild-flooding:* The water is released at high points in the field and distribution is uncontrolled.

Interstratification: Mixing of silicate layers within the structural framework of a given silicate clay.

Isomorphous substitution: The replacement of one atom by another of similar size in a crystal lattice without disrupting or changing the crystal structure of the mineral.

Isotopes: Two or more atoms of the same element that have different atomic masses because of different numbers of neutrons in the nucleus.

Jevenile: New or young animal or plant or soil.

Joule: The SI energy unit defined as a force of one newton applied over a distance of one meter, 1 joule = 0.239 calorie.

Kame: A conical hill or ridge of sand or gravel deposited in contact with glacial ice.

Kainite [$KMg (SO_4)Cl.3H_2O$] A mineral used as a source for potassic fertilizers and potassic salts. It is an impure mixture of potassium and magnesium chloride, sometimes containing sulphate of magnesia. It contains about 12 per cent K_2O. It is quite soluble in water and is non-hygroscopic.

Kalium: The Latin word for alkali. In chemistry it stands for potassium and the symbol 'K' used for potassium is derived from kalium.

Kaolinite: An aluminosilicate mineral of the 1: 1 crystal lattice group, that is, consisting of single silicon tetrahedral sheets alternating with single aluminium octahedral sheets.

Kinetic energy: Energy resulting from motion. In equation form, KE= ½ mass x (velocity squared).

Kjedahl nitrogen: The amount of nitrogen contained in organic material as determined by a method based on the digestion of the sample in a sulphuric acid base reagent, which converts the nitrogenous organic material to carbon dioxide, water and ammonia. Subsequently the ammonia is quantified.

Labile: Descriptive of a substance in soil that readily undergoes transformation or is readily available to plants.

Lacustrine deposit: Material deposited in lake water and later exposed either by lowering of the water level or by the elevation of the land.

Lambert's Law: The law states that light absorbed by homogeneous transparent media is proportional to the thickness of the media and intensity of light. Mathematically, Lambert's law is described as:

$$Log = \frac{o}{It} \, KH$$

Io = Intensity of incident light

It = Intensity of transmitted light

H = Thickness of media

K = Constant.

Land: A broad term embodying the total natural environmental of the areas of the Earth not covered by water. In addition to soil, its attributes include other physical conditions such as mineral deposits and water supply; location in relation to centers of commerce, populations, and other land, the size of the individual tracts or holdings, and existing plant cover, works of improvement, and the like.

Land capability classification: A grouping of kinds of soil into special units, subclasses, and classes according to their capability for intensive use and the treatments required for sustained use. One such system has been prepared by the USDA Soil Conservation Service.

Land classification: The arrangement of land units into various categories based upon the properties of the land or its suitability for some particular purpose.

Landslide: A rapid downhill movement of soil and rock under the force of gravity. A term also used to described the resulting landform.

Land-use planning: The development of plans for the uses of land that, over long periods, will best serve the general welfare, together with the formulation of ways and means for achieving such uses.

Latent heat: A property of a material. The amount of heat involved per unit mass when the material undergoes a change of phase (*e.g.*, the heat of melting or the heat of vaporization).

Laterite: An iron-rich subsoil layer found in some highly weathered humid tropical soils that, when exposed and allowed to dry, becomes very hard and will not soften when rewetted. When erosion removes the overlying layers, the laterite is exposed and a virtual pavement results.

Layer silicates: Another term for aluminosilicate minerals that are platelike in form and composed of layers of atoms. Mica and the aluminosilicate clay minerals are layer silicates.

Leaching: Removal of plant nutrients in solution by the passage of water through soil. This is one of the ways in which plant nutrients are lost from the soil. Among the major nutrients, nitrogen is lost in large quantities by leaching.

Leaf area index (LAI): Area of leaves per unit of land on which the plants are growing.

Legume Inoculation: Treatment of legume seed with rhizobium culture. There is a specific symbiotic relationship between different species of rhizobium and various legume crops. For example, a rhizobium specie that will live symbiotically with soyabean will not do so with alfalfa. It is, therefore, necessary to use only specific cultures for different crops.

Legume: A pod-bearing member of the leguninosae family, one of the most important and widely distributed plant families. Includes many valuable food and forage species, such as peas, beans, peanuts, clovers, alfalfas, sweet clovers, lespedezas, vetches, and kudzu. Nearly all legumes are associated with nitrogen-fixing organisms.

Lichen: Symbiosis between fungi and algae or bluegreen bacteria, commonly forming a flat, speading growth on surfaces of rocks and tree trunks.

Liebig's law: In 1862, Justus Von Liebing (A German Chemist) stated that the growth and reproduction of an organism are determined by the nutrient substance (oxygen, carbon dioxide, calcium, *etc*) that is available in minimum quantity, the limiting factor.

Ligand exchange: A class of surface reactions on minerals in which anions from solution take the place of some of the anions (ligands) normally bound to cations in the mineral.

Light soil: A term used for sandy or coarse-textured soil. Leaching of nutrients is more in light soils, as such it is advisable to apply nitrogenous fertilizers in small doses. Wherever practicable, green manuring should be done to improve the physical condition of a light soil.

Lignin: The complex organic constituent of woody fibers in plant tissue that, along with cellulose, cements the cells together and provides strength. Lignins resist microbial attack and after some modification may be come part of the soil organic matter.

Lime (agricultural): In strict chemical terms, calcium oxide. In practical terms, a material containing the carbonates, oxides and/or hydroxides of calcium and/or magnesium used to neutralize soil and acidity.

Lime: Generally, the term lime, or a agricultural lime, is applied to ground limestone, hydrated lime or burned lime. These materials are used as amendments to reduce the acidity of acid soils. In strict chemical terminology, lime refers to calcium oxide (CaO), but by an extension of the meaning, it is now used for all limestone-derived materials applied to neutralize acid soils.

Lime requirement: The amount of standard ground limestone required to bring a 6 inch layer of an acre (about two million pounds) of acid soil to some specific lesser degree of acidity. In common practice, lime requirements are given in tons per acre of nearly pure limestone.

Limestone: The term refers to rocks consisting chiefly of calcium carbonate ($CaCO_3$) or calcium and magnesium carbonate $CaMg(CO_3)_2$. When magnesium carbonate is present in a large percentage, limestone is called dolomite limestone.

Liming: Addition of ground limestone (calcium carbonate), hydrated lime (calcium hydroxide) or burned lime (calcium oxide) to acid soils to permit growth of deeply rooted legumes and many other valuable crops, and to create soil conditions favourable for the utilization of plant nutrients.

Liquid fertilizers: Commercial fertilizers in liquid form. Such fertilizers are chiefly anhydrous ammonia, aqueous solution of nitrogen, and some mixed fertilizers. Liquid fertilizers are applied to the soil, through irrigation water, starter solutions, or with the help of special equipment. Use of liquid fertilizers is common in the United States of America.

Liquid limit: In engineering, the water content corresponding to the limit between a soil's liquid and plastic states of consistency.

Lithosequence: Two or more soils in which the soil forming factor that varies the most is the parent material. Other soil-forming factors are constant or vary much less than parent material.

Loam: The textural class name for soil having a moderate amount of sand, silt, and clay. Loam soils contain 7-27 per cent clay, 28-50 per cent silt, and 23-52 per cent sand.

Loamy: Intermediate in texture and properties between fine-textured and coarse-textured soils. Includes all textural classes with the words loam or loamy as a part of the class name, such as clay loam or loamy sand.

Localized placement of fertilizers: This refers to the application of fertilizers into the soil close to the seed or plant. This method is specifically adopted with phosphatic fertilizers for three reasons, namely:

1. Restricted contact of fertilizers with soil lessens fixation of phosphate,
2. Necessary plant food is placed within easy reach of plant roots, and
3. Application of fertilizers in a band along the rows does not readily furnish nutrients to weeds growing between the rows.

Loess: Material transported and deposited by wind and consisting of predominantly silt-sized particles.

London forces: Weak, close-range attraction between atoms, molecules, or colloid particles due to transitory dipole-dipole interactions between their electrons.

Longwave radation: In meteorology, radiation in the infrared and radio wavelengths, emitted at the earth's fsurface and partly absorbed in the stomosphere. Distinct from the sun's shortwave radiation.

Low-analysis fertilizer materials: The fertilizers containing a low percentage of plant nutrients. Such fertilizers are Chilean nitrate containing 16 per cent nitrogen and single superphosphate containing 16 per cent nitrogen and single superphosphate containing 16 to 18 per cent P_2O_5.

Luxury consumption: The intake by a plant of an essential nutrient in amounts exceeding what is needs. For example, if potassium is abundant in the soil, alfalfa may take in more than it requires.

Lysimeter: A device for measuring perolation and leaching and evapotranspiration losses from a column of soil under controlled conditions.

Macronutrient: A nutrient required by plants in relatively large amounts. There are three macronutrients namely, nitrogen, phosphorus and potassium which are also called major nutrients or macro elements. Other nutrients which are normally required in lesser but still considerable amounts like calcium, magnesium and sulphur are called secondary nutrients. Some authors group secondary nutrients along with macronutrients.

Magnesium sulphate: ($MgSO_4, 7H_2O$) is commonly used to supply magnesium to the soil through fertilizer mixtures or foliar sprays.

Manure: The excreta of animals – dung and urine, with straw or other materials used as the absorbent. The decomposed manure is called farmyard manure or farm manure or barnyard manure. The average composition of well-rotted farmyard manure is 0.5 per cent nitrogen, 0.3 per cent P_2O_5 and 0.5 per cent K_2O.

Marl: Soft and unconsolidated calcium carbonate, usually mixed with vary8ing amounts of clay or other impurities.

Marsh: Periodically wet or continually flooded area with the surface not deeply submerged. Covered dominantly with sedges, cattails, rushes, or other hydrophytic plants. Subclasses include freshwater and saltwater marshes.

Marsh spot of pea: Disease caused in peas due to deficiency of manganese.

Mature soil: A soil with well developed soil horizons produced by the natural processes of soil formation and essentially in equibrium with its present environment.

Maximum water-holding capacity: The average moisture content of a disturbed sample of soil, 1 cm hgh, which is at equilibrium with a water table at its lower surface.

Medium texture: Intermediate between fine-textured and coarse textured (soils). It includes the following textural classes: very fine sandy loam, loam, silt loam, and silt.

Mellow Soil: A very soft, very friable, prorous soil without any tendency toward hardness or harshness.

Metamorphic rock: A rock that has been greatly altered from its previous condition through the combined action of heat and pressure. For example, marble is a metamorphic rock produced from limestone, gneiss is produced from granite, and slate from shale.

Metamorphic rock: Rocks which have undergone change in structure and mineral composition due to very high pressure and temperature conditions, *e.g.* when limestone is altered to marble.

Methane, CH_4: An odorless, colorless gas commonly produced under anaerobic conditions. When released to the upper atmosphere, methane contributes to global warming.

Micas: Primary aluminosilicate minerals in which two silica tetrahedral sheets alternate with one alumina/magnesia octahedral sheet with entrapped potassium atoms fitting between sheets. They separate readily into thin sheets or flakes.

Micron (μ): A unit of length denoting a millionth part of a metre.

$1 \mu = 10^{-6}$ metre.

In instrumental analysis it is used to measure wavelength of radiation.

Microfauna: That part of the animal population which consists of individuals too small to be clearly distinguished without the use of a microscope. Includes protozoans and nematodes.

Microflora: That part of the plant population which consists of individuals too small to be clearly distinguished without the use of a microscope. Includes actinomycetes, algae, bacteria, and fungi.

Micronutrients: The essential plant nutrients required in minute quantities. These nutrients are seven in number, namely, iron, manganese, boron, molybdenum, copper, zinc and chlorine. Micronutrients are also called 'minor elements' or 'trace elements'.

Micro-organisms: The most primitive plant and animal life whose structure is very simple. Their size being small, they are often found in soil and have direct or indirect bearing on soil formation and soil fertility. They are divided into two main groups:

1. Micro flora, *e.g.* bacteria, actinomycetes, fungi and algae.
2. Micro fauna, *e.g.* protozoa.

Microrelief: Small-scale local differences in topography, including mounds, swales, or pits that are only a meter or so in diameter and with elevation differences of up to 2 m.

Mineral soil: A soil consisting predominantly of, and having its properties determined predominantly by mineral matter. Usually contains <20 per cent organic matter, but may contain an organic surface layer upto 30 cm thick.

Mineralization: The conversion of an element from an organic form to an inorganic state as a result of microbial decomposition.

Mitscherlich's growth law: It states that increase in the yield of a crop as a result of increasing a single growth factor is proportional to the decrement from the maximum yield obtainable by increasing the particular growth factor.

Mixed fertilizers: Mixed fertilizers consist of individual or straight fertilizer materials blended together to permit application in the field in one operation. Mixed fertilizers supply two or three major plant nutrients. The percentage of these nutrients are expressed as fertilizer grade, like 10-5-5. See 'fertilizer grade' for further information.

Moderately coarse texture: Consisting predominantly of coarse particles. In soil textural classification, it includes all the sandy loams except the very fine sandy loam.

Moderately fine texture: Consisting predominantly of intermediate-sized (soil) particles or with relatively small amounts of fine or coarse particles. In soil textureal classification, it includes clay loam, sandy loam, sandy clay loam, and silty clay loam.

Moisture equivalent: The weight percentage of water retained by a previously saturated sample of soil 1 cm in thickness after it has been subjected to a centirifugal force of 1000 times gravity for 30 min.

Molar solution: One mole (gram molecular weight) of a substance dissolved in solvent and the fiinal volume made to one litre of solution, *e.g.* the gram molecular weight of sodium dydroxide is 40. Therefore a molar solution is prepared by dissolving 40 g of sodium hydroxide in water and making up the final volume of the solution (by adding water) to one litre.

Molar concentration refers to number of moles in one litre solution, *e.g.* 2 molar concentration of sodium hydroxide solution means 2 moles (40 x 2 = 80 g) of sodium hydroxide present in one litre of solution.

Mole drain: It is unlined drain formed by pulling a bullet-shaped cylinder through the soil.

Montmorillonite: An aluminosilicate clay mineral in the smectite group with a 2: 1 expanding crystal lattice, with two silicon tetrahedral sheets enclosing an aluminum octahedral sheet. Isomorphous substition of magnesium for some of the aluminum has occurred in the octahedral sheet. Considerable expansion may be caused by water moving between silica sheets of contiguous layers.

Mor raw: It is a type of forest humus layer of unincorporated organic material, usually matted or compacted or both, distinct from the mineral soil, unless the latter has been blackened by washing in organic matter.

Moraine: An accumulation of drift, with an initial topographic expression of its own, built within a glaciated region chiefly by the direct action of glacial ice. Examples are ground, lateral, recessional, and terminal moraines.

Morphology, soil: The constitution of the soil including the texture, structure, consistence, color, and other physical, chemical and biological properties of the various soil horizons that make up the soil profile.

Most profitable rate or MPR or optimum dose: Refers to the amount of fertilizer that produces the greatest profit per acre. If, more than the optimum dose is given, the profit will fall and less than the optimum is used, some profit will be sacrificed.

Motting: Spots or blotches of different color or shades of color interspersed with the dominant color.

Mucigel: The gelatinous material at the surface of roots grown in unsterilized soil.

Muck: Highly decomposed organic material in which the original plant parts are not recognizable. Contains more mineral matter and is usually darker in color than peat.

Muck soil: (1) A soil containing 20-50 per cent organic matter. (2) An organic soil in which the organic matter is well decomposed.

Mull: A humus-rich layer of forested soils consisting of mixed organic and mineral matter. A mull blends into the upper mineral layers without an abrupt change in soil characteristics.

Multiple nutrient materials: The fertilizers containing more than one nutrient element. When fertilizers supply two major plant nutrients they are called binary fertilizers. Ammonium phosphate and nitrophosphate are binary fertilizers. Fertilizers supplying three major plant nutrients are termed ternary fertilizer. Ammonium potassium phosphate is a ternary fertilizer.

Muriate of potash: It is potassium chloride, of 90 to 99 per cent grade. It contains nearly 60 per cent K_2O. It is sold in granular as well as powder form.

Mycorrhiza: A mycorrhiza is an infected root system arising from the root lets of a seed plant. The word mycorrhiza (my-koe-rye-zee), derived from Greek meaning "fungus root" Mycorrhizae are fungi that form symbiotic association of a fungus with the roots of a higher plants.

Necrosis: Death associated with discoloration and dehydration of all or parts of plant organs, such as leaves.

Neem coated urea: Neem has a nitrification inhibition property, hence is used in India for coating of urea granules to produce slow release fertilizer. The technique as developed at the Indian Agriculture Research Institute for coating urea with neem cake is as follows: About 100 kg urea is mixed in a drum with a solution of 1 kg coaltar in 2 litres of kerosene. To this, 20 kg of powdered neem cake is added and thoroughly mixed.

Nematodes: Very small worms abundant in many soils and important because some of them attack and destroy plant roots.

Neutral soil: A soil in which the surface layer, at least to normal plow depth, is neither acid nor alkaline in reaction. In practice this means the soil is within the pH range of 6.6 – 7.3.

Night soil: Night soil is human excreta, solid and liquid, In India, it is directly applied to the soil to a limited extent but converted mainly as town compost. In cities which have sewage facilities, sewage water and sludge are used directly to raise crops. On an average nigh-soil contains 5.5 per cent nitrogen, 4.0 per cent phosphorus (P_2O_5) and 2.0 per cent potash (K_2O) on oven dry basis.

Niter (KNO_3): Potassium bearing mineral also known as potassium nitrate or saltpeter. It contains about 46 per cent K_2O.

Nitrate of soda: A commercial nitrogenous fertilizer carrying 16 per cent nitrogen in nitrate form. Nitrate of soda is imported from Chile hence the name Chilean nitrate. In other countries, it is manufactured synthetically.

Nitrification: The formation of nitrates and nitrites from ammonia (or ammonium compounds) as in soils, by micro-organisms.

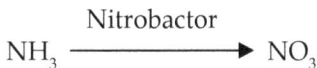

$$NH_3 \xrightarrow{\text{Nitrosomonas}} NO_2 + 3 H_+$$

$$NH_3 \xrightarrow{\text{Nitrobactor}} NO_3$$

Nitrification inhibitors: Chemicals which inhibits the nitrification process, used extensively to check mineralization of nitrogen and thereby control the release of nitrogen. Some of the important nitrification inhibitors are N-serve, AM, DD, Thiourea and ST.

Nitrogen cyle: The sequence of chemical and biological changes undergone by nitrogen as it moves from the atmosphere into water, soil, and living organisms, and upon death of these organisms (plants and animals) is recycled through a part or all of the entire process.

Nitrogen fixation: The assimilation of free nitrogen from the soil air by soil micro-organisms and the formation of nitrogen compounds that eventually become available to plants. The nitrogen fixing organisms associated with legumes are called symbiotic; those not definitely associated with higher plants are non-symbiotic.

Nitrophosphate: This fertilizer contains 20 per cent phosphorus. Nitrophosphate is manufactured by treating rock phosphate with nitric acid and neutralizing this solution with gaseous ammonia. The nitrogen is present in ammonium nitrate form, and P_2O_5 partly in water-soluble and partly in citrate-soluble form.

Nucleic acids: Complex compounds found in plant and animal cells may be combined with proteins as nucleoproteins.

Nutrient deficiency symptoms: When any essential plant nutrient is seriously lacking in the soil, plants growing on it show certain colour development in leaves and certain changes in growth. These symptoms varies from crop to crop and with the degree of deficiency. On the basis of deficiency symptoms, fertilizer recommendations can be done.

Nutrient: Any mineral element that functions in plant, animal and other organism, metabolisms, whether or not its action is specific.

O horizon: Organic horizon of mineral soils.

Oilcakes: When oil is extracted from oilseeds, the remaining solid portion is the oilcake. Oilcakes are of two types, edible oilcakes which can be safely fed to livestock, and non-edible oilcakes which are not fit for feeding to livestock. Oilcakes are added to the soil as concentrated organic manures. They supply organic matter and all the three major plant nutrients (N, P_2O_5, and K_2O) but mostly they supply nitrogen.

Open-formula mixed fertilizers: Fertilizer mixtures of which the composition and the complete make up are disclosed by manufacturers. Thus the fertilizer grade, the straight fertilizers and the filler used are disclosed on each bag. Cultivators and extension workers are in a better position to know the quantity and quality of nutrients contained in such a fertilizer mixture.

Organic colloids: In soils, these present as humus which is a temporary intermediate product left after considerable decomposition of plant and animal residues. Organic colloids having high surface area with higher charge density and higher cation exchange capacity.

Organic farming: According to the U.S. Department of Agriculture "a production system which avoids or largely excludes the use of synthetically compounded fertilizers, pesticides, growth regulators and livestock feed additives. To the maximum extent feasible, organic forming systems rely upon crop rotations, crop resideues, animal manures, legumes, green manures, off-farm organic waste, mechanical cultivation, mineral bearing rocks and aspects of biological pest control to maintain soil productivity and tilth, to supply plant nutrients and to control insects, weeds and other pests".

Organic fertilizer: By product from the processing of animal or vegetable substances that contain sufficient plant nutrients to be of value as fertilizers.

Organic soil: A soil that contains at least 20 per cent organic matter (by weight) if the clay content is low and at least 30 per cent if the clay content is as high as 60 per cent.

Organic soil materials (As used in Soil taxonomy in the United States). (1) Saturated with water for prolonged periods unless artificially drained and having 18 per cent or more organic carbon (by weight) if the mineral fraction is more than 60 per cent clay, more than 12 per cent, organic carbon if the mineral fraction has no clay, or between 12 and 18 per cent carbon if the clay content of the mineral fraction is between 0 and 60 per cent. (2) Never saturated with water for more than a few days and having more than 20 per cent organic carbon. Histosols develop on these organic soil materials.

Organic matter: The terms organic means "living", therefore, organic matter is any material derived from living organism like plants and animals. The organic matter content in soil is an important indicator of nutrient status especially of nitrogen content in soils. Generally, the status of soil organic carbon content less than 0.5 per cent is termed as low and more than 0.75 per cent as high.

Ortstein: An indurated layer in the B horizon of spodosols in which the cementing material consists of illuviated sesquioxides (mostly iron) and organic matter.

Osmotic pressure: Pressure exerted in living bodies as a result of unequal concentrations of salts on both sides of a cell wall or membrane. Water moves from the area having the lower salt concentration through the membrane into the area having the higher salt concentration and, therefore, exerts additional pressure on the side with higher salt concentration.

Outwash plain: A deposit of coarse-textured materials (*e.g.*, sands, gravels) left by streams of melt water flowing from receding glaciers.

Oven-dry soil: Soil that has been dried at 105°C until it reaches constant weight.

Oxidation ditch: An artificial open channel for partial digestion of liquid organic wastes in which the wastes are circulated and aerated by a mechanical device.

Oxisols: A soil order which is highly weathered dominant in oxides of iron and aluminium but depleted of silica. Oxisols are reddish, yellowish or greyish soils of tropical and sub-tropical regions and low in fertility.

Pans: Horizons or layers, in soils, that are strongly compacted, indurated, or very high in clay content.

Parent material: The unconsolidated and more or less chemically weathered mineral or organic matter from which the solemn of soils is developed by pedogenic processes.

Particle density: The mass per unit volume of the soil particles. In technical work, usually expressed as metric tons per cubic meter (Mg/m^3) grams per cubic centimeter (g/cm^3).

Particle size: The effective diameter of a particle measured by sedimentation, sieving, or micro-metric methods.

Particle-size analysis: Determination of the various amounts of the different separates in a soil sample, usually by sedimentation, sieving, micrometry, or combinations of these methods.

Particle-size distribution: The amounts of the various soil separates in a soil sample, usually expressed as weight percentages.

Pascal: An SI unit of pressure equal to one newton per square meter.

Peat: Unconsolidated soil material consisting largely of undecomposed, or only slightly decomposed, organic matter accumulated under conditions of excessive moisture.

Ped: A unit of soil structure such as an aggregate, crumb, prism, block, or granule, formed by natural processes (in contrast to a clod, which is formed artificially).

Pedon: The smallest volume that can be called "a soil". It has three dimensions. It extends downward to the depth of plant roots or to the lower limit of the genetic soil horizons. Its lateral cross section is roughly hexagonal and ranges from 1 to 10 m^2 in size depending on the variability in the horizons.

Pedology: It is branch of soil science which deals with the survey, genesis and classification of soils for land use planning.

Petrology: It is science of rocks which form the units of the earth's crust.

Peneplain: A once high, rugged area that has been reduced by erosion to a lower, gently rolling surface resembling a plain.

Penetrability: The ease with which a probe can be pushed into the soil. May be expressed in units of distance speed, force, or work depending on the type of penetrometer used.

Percolation, soil water: The downward movement of water through soil. Especially, the downward flow of water in saturated or nearly saturated soil at hydraulic gradients of the order of 1.0 or less.

Permafrost: (1) Permanently frozen material underlying the solum. (2) A perennially frozen soil horizon.

Permeability, soil: The ease with which gases, liquids, or plant roots penetrate or pass through a bulk mass of soil or a layer of soil.

pH: A term used to indicate the degree of acidity or alkalinity, Technically, pH is the common logarithm of the reciprocal of the hydrogen ion concentration of a solution. A pH of 7.0 indicates precise neutrality, higher values indicate increasing alkalinity, and lower values indicate increasing acidity.

Phase, soil: A subdivision of a soil series or other unit of classification having characteristics that affect the use and management of the soil but do not vary sufficiently to differentiate it as a separate series. Included are such characteristics as degree of slope, degree of erosion, and content of stones.

pH-dependent charge: That portion of the total charge of the soil particles that is affected by, and varies with, changes in pH.

Phosphobacterium: A microbial culture containing bacteria which increases the availability of applied and native soil phosphorus. Several species of bacteria are effective for increasing phosphorus availability but *Bacillus magatherium var phosphaticum* is most commonly used. In Russia and several Eastern European countries phosphobacterium culture is sold commercially and is used to inoculate the soil for increasing phosphorus availability.

Phosphoric acid: A term that refers to the phosphorus content of a fertilizer, expressed as phosphoric acid (P_2O_5).

Photomap: A mosaic map made from aerial photographs to which place names, marginal data, and other map information have been added.

Phyllosphere: The leaf surface.

Physical properties (of soils): Those characteristics, processes, or reactions of a soil that are caused by physical forces and that can be described by or expressed in, physical terms or equations. Examples of physical properties are bulk density, water holding capacity, hydraulic conductivity, porosity, pore-size distribution, and so on.

Plant tissue test: This refers to the rapid chemical testing of plant tissue, which is analogous to chemical testing and diagnosis in medicine. The basis of rapid chemical testing of plant tissue is the colorimetric determination of the levels of nitrate, phosphorus and potassium in the sap of fresh plant tissue. Sometimes, tests are also carried out for magnesium, calcium, manganese and zinc. Plant tissue tests, though only qualitative, are highly useful guides to interpretation of the relative supply of nutrients actually being taken by the plant. The underlying assumption is that an adequate supply of the elements is indicated by abundance in the plant sap.

Plastic soil: A soil capable of being molded or deformed continuously and permanently, by relatively moderate pressure, into various shapes.

Platy: Consisting of soil aggregates that are developed predominantly along the horizontal axes, laminated, flaky.

Plinthite (brick): A highly weathered mixture of sesquioxides or iron and aluminum with quartz and other diluents that occurs as red mottles and that changes irreversibly to hardpan upon alternate wetting and drying.

Plough-sole placement: With the plough-sole method, the fertilizer is placed in a continuous band on the bottom of a furrow in the process of ploughing. Each band is covered as the next furrow is turned. This method is advocated in areas where the soil becomes quite dry for a few centimeters below the surface during the growing season.

Plow layer: The soil ordinarily moved when land is plowed, equivalent to surface soil.

Plow pan: A subsurface soil layer having a higher bulk density and lower total porosity than layers above or below it, as a result of pressure applied by normal plowing and other tillage operations.

Plowing: A primary broad base tillage operation that is performed to shatter soil uniformly with partial to complete inversion.

Polynutrient fertilizer: A fertilizer containing more than one major plant nutrient. It is synonymous for multiple nutrient material and complex fertilizers.

Polypedon (as used in soil taxonomy): Two or more contiguous pedons, all of which are within the defined limits of a single soil series, commonly referred to as a soil individual.

Pore size distribution: The volume of the various sizes of pores in a soil. Expressed as percentages of the bulk volume (soil plus pore space).

Porosity, soil: The volume percentage of the total soil bulk not occupied by solid particles.

Potash (potassium oxide): The potassium content of fertilizer is expressed as K_2O.

Potassium chloride: Commercial potassium chloride is a potash salt that contains not less than 48 per cent potash, chiefly as chloride, Normally, potassium chloride or muriate of potash sold in India supplies 60 per cent K_2O.

Potassium sulphate: Commercial potassium sulphate contains not less than 48 per cent potash (K_2O) chiefly as sulphate, and not more than 20.5 per cent as chloride.

Precision agriculture: It is the application of modern information technologies and makes uses of information system and planning software to provide, process and analyze multi-source data of high spatial and temporal resolution for decision making and operation in the management of crop production.

Primary mineral: A mineral that has not been altered chemically since deposition and crystallization from molten lava.

Primary plant foods: The primary plant foods are nitrogen (N)., phosphoric acid (P_2O_5) and potash (K_2O). These nutrients are also called major plant nutrients.

Prismatic soil structure: A soil structure type with prism-like aggregates that have a vertical axis much longer than the horizontal axes.

Productivity, soil: The capacity of a soil for producing a specified plant or sequence of plants under a specified system of management. Productivity emphasizes the capacity of soil to produce crops and should be expressed in terms of yields.

Profile, soil: A vertical section of the soil through all its horizons and extending into the parent material.

Protein: Any of a group of nitrogen-containing organic compounds formed by the polymerization of a large number of amino acid molecules and that, upon hydrolysis, yield these amino acids. They are essential parts of living matter and are one of the essential food substances of animals.

Puddled soil: Dense, massive soil artificially compacted when wet and having no aggregated structure. The condition commonly results from the tillage of a clayey soil when it is wet.

Putrefaction: A process of degradation by which protein rich material is decomposed under anaerobic conditions thereby releasing foul smelling gases. The final products in putrefaction are ammonia, amines, carbon dioxide, organic acids, hydrogen sulphide *etc.*

Quartz (SiO_2): Hard crystalline silica found extensively in igneous and sedimentary rocks. The sand used as filler in fertilizers has quartz as the main constituent.

Quantitative chemical analysis: The chemical testing of a substance to determine its chemical nature. It identifies chemical substance in a product.

Quick soil tests: These are simple and rapid chemical tests of soils designed to give an approximation of the nutrients available to plants. Interpretation of results depends upon previous standardization of soil tests with responses to fertilizers. It varies for different kinds of soil.

Radiation: Processes in which energy is sent out as waves and particles through space from atoms and molecules as they rotate, vibrate, and undergo internal change.

Rangeland: Land used for free-grazing livestock.

Reagent: Any chemical compound used to produce a chemical reaction in laboratory analysis to detect and identify specific constituents of a substance being examined.

Recalcitrant: The property of a material to stub-bornly resist decomposition. A chemical such as a pesticide that is not decomposed by microorganisms is a recalcitrant pesticide.

Reclamation: In regard to soil, the treatment of soil to correct waterlogging or severe excesses such as salinity or sodicity.

Reduction: Addition of electron(s) to an atom or molecule. Opposite of *axidation.*

Reemitted energy: Radiation of previously absorbed energy.

Reflected energy: Radiant energy that is thrown back from an object, with no change except direction.

Regolith: The unconsolidated mantle of weathered rock and soil material on the earth's surface, loose earth materials above solid rock. (Approximately equivalent to the term "soil" as used by many engineers).

Relative humidity: The concentration of water in the air or soil atmosphere relative to the maximum concentration it can hold at the given temperature.

Remote sensing: Identifying and observing objects without having any contact with the object. It is feasible through the use of aerial photographs or setellite imagery.

Reradiation: Radiant emission of energy previously absorbed.

Residual effect of fertilizers: This refers to the amount of fertilizer that remains in the soil after one or more cropping seasons. This is particularly true of phosphatic fertilizers.

Residual effect of manure: This refers to the residual beneficial effect of application of farmyard manure on the succeeding crops. This beneficial effect is due to improvement in the physical condition of the soil, and also due to the unutilized plant nutrients. It is estimated that only one third of the nitrogen present in farmyard manure is utilized by the first crop. Similarly, about two third of the phosphate is effective but most of the potash is available for the first crop.

Residual sodium carbonate (RSC): It is a parameter to assess the quality of irrigation water can be expressed by following formula.

$$RSC = (CO_3^{2-} + HCO_3^-) - (Ca^{2+} + Mg^{2+})$$

All concentrations are expressed in me/L.

Residual material: Unconsolidated and partly weathered mineral materials accumulated by disintegration of consolidated rock in place.

Residual parent material: Unconsolidated and partly weathered mineral materials accumulated by disintegration of consolidated rock in place.

Residue conservation: Leaving straw, stubble, trash – crop residues – to rot on or in the ground instead of removing or burning them.

Resistance: A measure of the difficulty with which a material conducts heat, ions, electricity, or water. Reciprocal of *conductance.*

Respiration: The set of metabolic processes in which an organism obtains energy from the oxidation of sugars to CO_2 and water, usually with O_2 as the oxidizing agent.

Response curve: A graph relating a plant's quantitative response-changes in growth, yield, or any desired attribute-to different levels o a treatment such as fertilizer application.

Rhizobia: Bacteria capable of living symbiotically with higher plants, usually in nodules on the roots of legumes, from which they receive their energy, and capable of converting atmospheric nitrogen to combined organic forms; hence, the term *symbiotic nitrogen-fixing bacteria.* (Derived from the generic name *Rhizobium).*

Rhizosphere: The term was introduced by L. Hiltner (1904) to describe that portion of the soil which is in close contact with roots and is influenced by the root system with regard to microbial activity and other related phenomena.

Rill: A small, intermittent water course with steep sides, usually only a few centimeters deep and hence no obstacle to tillage operations.

Ripening: The process by which reclaimed land develops characteristics suitable for farming. It includes drainage, cultivation, and neutralizing acidity.

Riprap: Broken rock, cobbles or boulders placed on earth surfaces, such as the face of a dam or the bank of a stream, for protection against the action of water (waves); also applied to brush or pole mattresses, or brush and stone, or other similar materials used for soil erosion control.

Rock: The material that forms the essential part of the Earth's solid crust, including loose incoherent masses such as sand and gravel, as well as solid masses of granite and limestone.

Rock phosphate: Rock containing phos-phorus such as tricalcium phosphate. Rock phosphate is treated with strong acids to make superphosphate. Finely ground rock phosphate is used in strongly acid soils to supply P_2O_5. The finer the rock phosphate, more effective it will be on acid soils.

Root cap: Cells at the tip of the root protecting the growing cells as the root pushes through the soil.

Root density: Number mass or length of roots within a given soil volume.

Root distribution: Arrangement of roots in space within the soil, especially the vertical root density profile.

Root exudate: A mixture of organic acids, sugars, and other soluble plant components that escape from roots.

Root hair: A hairlike structure produced from a root epidermal cell. Probably helps obtain water and immobile nutrients.

Rotational slide: Mass wasting in which the mass rotates about a point rather than sliding horizontally.

Runoff: The portion of the precipitation on an area that is discharged from the area through stream channels. That which is lost without entering the soil is called surface runoff and that which enters the soil before reaching the stream is called *ground water runoff or seepage flow* from groundwater. (In soil science "runoff" usually refers to the water lost by surface flow; in geology and hydraulics "runoff" usually includes both surface and subsurface flow).

Saline: Containing large concentrations of soluble salts. Operationally defined as the electrical conductivity of a saturation extract of > 4 deciSiemens per meter (dS m^{-1})

Saline soil: A saline soil contains enough soluble salts so distributed in the soil that they interfere with growth of most crop plants. Ordinarily the saline soil is only slightly alkaline in reaction (pH 7.4 to 8.5) and contains very little absorbed sodium. Saline soils are often recognized by the presence of white salt crusts. As such, a saline soil is sometimes referred to as white alkali soil.

Salinization: The process of accumulation of salts in soil.

Saltation: Particle movement in water or wind where particles skip or bounce along the stream bed or soil surface.

Salt index: It is one of criteria for evaluating the quality of an irrigation water. It can be expressed as follows:

SI = (Total Na-24.5) − [Total Ca − Ca in $CaCO_3$) x 4.85].

Sand: Soil of particle size ranging from 0.02 to 2 mm in diameter formed as a result of the weathering of rocks. Sand is one of the fillers commonly used in fertilizer mixtures.

Saturated: Generally, occupying all of a capacity. With respect to water, the condition of a soil when all pores are filled with water. With respect to a particular cation or group of cations, the condition in which cations of the specified kind occupy all the exchange capacity.

Saturation extract: The solution extracted from a saturated soil paste.

Saturation percentage: The water content of a saturated soil paste, expressed as a dry weight percentage.

Savanna mineral: A mineral resulting from the decomposition of a primary mineral or from the repreipitation of the products of decomposition of a primary mineral.

Seal: A thin, dense layer that develops when the soil is wet.

Second bottom: The first terrace above the normal flood-plain of a stream.

Secondary plant nutrients: The secondary plant nutrients are calcium, magnesium and sulphur. These nutrients are called secondary because these are not applied as straightcommercial fertilizer but are applied to the soil, indirectly while adding N, P_2O_5 and K_2O in the form of commercial fertilizers. Thus for manufacturers of fertilizers containing major plant nutrients, calcium, magnesium and sulphur are of secondary importance.

Section-range system: A legal survey system used to accurately locate a parcel of land.

Sedimentary rock: A rock fo4med from materials deposited from suspension or precipitated from solution and usually being more or less consolidated. The principal sedimentary rock are sandstones, shales, limestones, and conglomerates.

Sedimentation: A technique for determining particle size distribution. Particles fall through water at a rate proportional to their size.

Seedbed: The soil prepared to promote the germination of seed and the growth of seedlings.

Self-mulching soil: A soil in which the surface layer becomes so well aggregated that it does not crust and seal under the impact of rain but instead serves as a surface mulch upon drying.

Semiarid: Term applied to regions or climates where moisture s more

Sensible heat: Ordinary everyday heat, sensed by feel or by thermometer.

Separate, soil: One of the individual-sized groups of mineral soil particles – sand, silt, or clay.

Septic tank: An underground tank used in the deposition of domestic wastes. Organic matter decomposes in the tank, and the effluent is drained into the surrounding soil.

Sesquioxides: The general term for the aluminum and iron oxides in soil.

Seventh Approximation: The first widely distributed version of the present U.S. soil classification system.

Sewage sludge: Settled sewage solids combined with varying amounts of water and dissolved materials, removed from sewage by screening, sedimentation, chemical precipitation, or bacterial digestion.

Shear: Force, as of a tillage implement, acting at right angles to the direction of movement.

Shear strength: The resistance of a soil to forces acting at right angles to the soil body. A plow produces shearing forces.

Sheet (mineralogy): A flat array of more than one atomic thickness and composed of one or more levels of linked coordination polyhedra. A sheet is thicker than a plane and thinner than a layer. Example, tetrahhedral sheet, octahedral sheet.

Sheet erosion: Water erosion that removes a uniform layer of soil from the land surface.

Sheet flow: A thin, relatively uniform water runoff (a few millimeters deep), in which the water is not concentrated in channels.

Shelterbelt: A wind barrier of living trees and shrubs established and maintained for protection of farm fields. Syn. Windbreak.

Shifting cultivation: A farming system in which land is cleared, the debris burned, and crops grown for 2-3 years. When the farmer moves on to another plot, the land is then left idle for 5-15 years, then the burning and planting process is repeated.

Shortwave radiation: In meteorology, high-energy radiation of the ultraviolet, visible, and near infrared wavelengths, sunshine. Unlike longwave radiation, it passes freely through air.

Shrinkage limit: The maximum water content at which a reduction in water content will not cause a decrease in the soil's volume.

Side dressing: The application of fertilizer alongside row crop plants, usually on the soil surface. Nitrogen materials are most commonly side-dressed.

Silica/alumina ratio: The molecules of silicon dioxide (SiO_2) per molecule of aluminum oxide (Al_2O_3) plus ferric oxide (Fe_2O_3) in clay minerals or in soils.

Silicates: Those mineals in which silicon and oxygen are the major elemental constituents.

Silt: 1) A soil separate consisting of particles between 0.05 and 0.002 mm in equivalent diameter. 2) A soil textural class.

Silting: The deposition of water-borne sediments in stream channels, lakes, reservoirs, or on floodplains, usually resulting from a decrease in the velocity of the water.

Sink: Somewhere or something for the disposal of wastewater.

Site index: A quantitative evaluation of the productivity of a soil for forest growth under the existing or specified environment.

Size separate: Individual mineral particles less than 2.0 mm in diameter, ranging between specified size limits. Sand, silt, and clay are common soil size separates.

Slag: A product of smelting, containing mostly silicates, the substances not sought to be produced as matte or metal and having a lower specific gravity.

Slake: A process of aggregate breakdown caused by internal pressures generated as water enters aggregates and air escapes.

Slash spots: Small areas in a field field that are slich when wet because of a high content of alkali or exchangeable sodium.

Slope: The degree of deviation of a surface from horizontal, measured in a numerical ratio, percent, or degrees.

Sludge: The solid portion of sewage. The sludge is obtained by treating sewage by different methods. Accordingly, sludges of different types are formed. Various types of sludges are settled sludge, digested sludge, activated sludge, digested activatedsludge and chemically precipitated sludge, On an average, sludge contains 1.5 to 3.5 per cent nitrogen, 0.75 to 4.0 per cent P_2O_5 and 0.3 to 0.6 per K_2O.

Slump: Collapse of soil and rock due to gravity.

Smectite: A group of silicate clays having a 2: 1 type lattice structure with suffcent isomorphous substitution in either or both the tetrahedral and octahedral sheets to give a high interlayer negative charge and high cation exchange capacity and to permit significant interlayer expansion and consequent shrinking and swelling of the clay. Montmorillonite, beidellite, and saponite are in the smectite group.

Sodic soil: A soil containing sufficient exchangeable sodium to interfere with the growth of most crop plants and containing appreciable quantities of soluble salts. The exchangeable sodium adsorption ratio is > 13, the conductivity of the saturation extract is >4 dS/m (at 25°C), and the pH is usually 8.5 or less in the saturated soil.

Sodium (Na): It is one of the most common alkali metals. Its atomic number is 11 and atomic weight 22.9 per cent. It forms a number of chemical compounds used very widely in industry.

It is found in plant ash but is not yet accepted as an essential element. However, it is known to have influence on plant cell osmotic pressure and metabolism.

From the angle of agriculture, sodium is useful for the following purposes:

1. It is an essential element for certain plants like sugar beet.
2. It acts as a substitute for potassium in certain plant metabolic functions.
3. It acts as an agent to maintain a high degree of availability of phosphates applied to the soil as superphosphate.

Its excess in soil is harmful for growth of plants. It is a dominanating cation in alkali soils.

Sodium adsorption ratio (SAR): The concentration of Na divided by the square root of the sum of Ca and Mg concentrations, both expressed as molarities and measured in the saturation extract.

$$SAR = \frac{Na^+}{\sqrt{\dfrac{Ca^{2+} + Mg^{2+}}{2}}}$$

Soil acidity factors: A set of related factors often inhibiting plant growth in acidic soils (deficiency of Ca^{2+} and excess of H^+, Al^{3+}, and Mn^{2+}).

Soil air: The soil atmosphere; the gaseous phase of the soil, being that volume not occupied by soil or liquid.

Soil alkalinity: The degree or intensity of alkalinity of a soil, expressed by a value > 7.0 on the pH scale.

Soil amendment: Any material, such as lime, gypsum, sawdust, or synthetic conditioner, that is worked into the soil to make it more amenable to plant growth.

Soil association: A group of defined and named taxonomic soil units occurring together in an individual and characteristic pattern over a geographic region, comparable to plant associations in many ways.

Soil classification (Soil Taxonomy): The systematic arrangement of soils into groups or categories on the basis of their characteristics.

Soil complex: A mapping unit used in detailed soil surveys where two or more defined taxonomic units are so intimately intermixed geographically that it is undesirable or impractical, because of the scale being used, to separate them. A more intimate mixing of smaller areas of individual taxonomic units than that described under *soil association*.

Soil conditioners: Are chemicals which are added to maintain physical condition of the soil. Such chemicals are polyvinytites polyacrylates, cellulose gums, lignin derivatives and silicates.

Soil conservation: A combination of all management and land-use methods that safeguard the soil against depletion or deterioriation caused by nature and/or humans.

Soil correlation: The process of defining, mapping, naming, and classifying the kinds of soils in a specific soil syrvey area, the purpose being to ensure that soils are adequately defined, accurately mapped and uniformly named.

Soil density: A measure of the relative amount of pores and solid particles. The mass per unit volume.

Soil drainage class: An interpretive grouping of soils based on the level of the water table during the growing season and the rate of water flow through soil.

Soil fabric: The pattern resulting from the arrangement of solids and pores.

Soil fertility: Soil fertility refers to the inherent capacity of a soil to supply nutrients to plants in adequate amount and in suitable proportion.

Soil forming factors: The five interrelated natural factors that are active in the formation of soil: parent material, climate, organisms, topography, and time.

Soil forming process: The biological, chemical and physical processes that act under the influence of the soil forming factors to create soils.

Soil genesis: The mode of origin of the soil, with special reference to the processes or soil-forming factors responsible for the development of the solum, or true soil, from the unconsolidated parent material.

Soil geography: A subspecialization of physical geography concerned with the areal distributions of soil types.

Soil individual: The pedon; a three-dimensional body large enough to contain all the properties necessary to describe a soil completely.

Soil management: The sum total of all tillage operations, cropping practices, fertilizer, lime, and other treatments conducted on or appliedto a soil for the production of plants.

Soil map: A map showing the distribution of soil types or other soil mapping units in relation to the promiment physical and cultural features of the Earth's surface.

Soil matrix: Like soil fabric, the combination of solids and pores in a soil.

Soil mechanics and engineering: A subspecialization of soil science concerned with the effect of forces on the soil and the application of engineering principles to problems involving the soil.

Soil monolith: A vertical section of a soil profile removed from the soil and mounted for display or study.

Soil morphology: The physical constitution, particularly the structural properties, of a soil profile as exhibited by the kinds, thicknesses, and arrangement of the horizons in the profile, and by the texture, structure, consistence, and porosity of each horizon.

Soil order: In soil classification, the most general level of classification. All soils fit into 12 orders.

Soil organic matter: The organic fraction of the soil that includes plant and animal residues at various stages of decomposition, cells and tissues of soil organisms, and substances synthesized by the soil population. Commonly determined as the amount of organic material contained in a soil sample passed through a 2 mm sieve.

Soil pH: It refers to reaction of the soil as expressed in terms of pH. Soil having pH 7 is neutral and with less than pH 7 is acidic. If the soil pH is more than 7, it is alkaline in reaction.

Soil potential: The usefulness of a site for a specific purpose using available technology at a cost expressed in economic, social, or environmental units of value.

Soil productivity: Productivity is the present capacity of a soil to produce crop yield under a defined set of management practices. It is measured in terms of the yield in relation to the input of production factors.

Soil profile: A vertical section of the soil from the surface through all its horizons, including C horizons.

Soil reaction: The degree of acidity of alkalinity of a soil mass, expressed in either pH value or in words

Soil reaction		*pH*
Extremely acid	-	Below 4.5
Very strongly acid	-	4.5 – 5.0
Strongly acid	-	5.1 – 5.5
Medium acid	-	5.6 – 6.0
Slightly acid	-	6.1 – 6.5
Neutral	-	6.6 – 7.4
Mildly alkaline	-	7.4 – 7.8
Moderately alkaline	-	7.9 – 8.4
Strongly alkaline	-	8.5 – 9.0
Very strongly alkaline		9.1 and higher

Soil salinity: The amount of soluble salts in a soil, expressed in terms of percentage, milligrams per kilogram, parts per million (ppm), or other convenient ratios.

Soil solution: The aqueous liquid phase of the soil and its solutes consisting of ions dissociated from the surfaces of the soil particles and of other soluble materials.

Soil structure: The combination or arrangement of primary soil particles into secondary particles, units, or peds. These secondary units may be but usually are not, arranged in the profile in such a manner as to give a distinctive characteristic pattern. The secondary units are characterized and classified on the basis of size, shape, and degree of distinctness into classes, types, and grades, respectively.

Soil structure classes: A grouping of soil structural units or peds on the basis of size from the very fine to very coarse.

Soil structure grades: A grouping or classification of soil structure on the basis of inter- and intra-aggregate adhesion, cohesion, or stability within the profile. Four grades of structure, designated from 0 to 3, are recognized.

0: Structureless – no observable aggregation.

1: Weakly durable peds.

2: moderately durable peds.

3: Strong, durable peds.

Soil structure types: A classification of soil stucture based on the shape of the aggregates or peds and their arrangement in the profile, including platy, prismatic, columnar, blocky, subangular blocky, granulated, and crumb.

Soil survey: The systematic examination, description, classification, and mapping of soils in an area. Soil surveys are classified according to the kind and intensity of field examination.

Soil taxonomy: The U.S. Department of Agriculture system of soil classification.

Soil temperature classes: Classes are based on mean annual soil temperature and on differences between summer and winter temperatures at a depth of 50 cm.

1. Soils with 5°C and greater difference between summer and winter temperatures are classed on the basis of mean annual temperatures.

   Frigid: < 8°C mean annual temperature.

   Mesic: 8-15 °C mean annual temperature.

   Thermic: 15-22 °C mean annual temperature.

   Hyperthermic: > 22°C mean annual temperature.

2. Soils with < 5°C difference between summer and winter temperatures are classed on the basis of mean annual temperatures.

   Isofrigid: < 8°C mean annual temperature.

   Isomesic: 8-15 °C mean annual temperature.

   Isothermic: 15-22 °C mean annual temperature.

   Isohyperthermic: <22°C mean annual temperature.

Soil-testing: This refers to chemical tests of the soil that can be made rapidly and with low cost, as compared to conventional methods of soil chemical analysis which are more accurate but more time-consuming and expensive. Soil-testing covers both rapid analysis in the field and in the laboratory.

Soil textural class: A grouping of soil textural units based on the relative proportions of the various soil separates (sand, silt, and clay). These textural classes, listed from the coarsest to the finest in texture, are sand, loamy sand, sandy loam, loam, silt loam, silt, sandy clay loam, clay loam, silty clay loam, sandy clay, silty clay, and clay. There are several subclasses of the sand, loamy sand, and sandy loam, classes based on the dominant particle size of the sand fraction (*e.g.,* loamy fine sand, coarse sandy loam).

Soil texture: The relative proportions of the various soil separates in a soil.

Soil water potential (total): A measure of the difference between the free energy state of soil water and that of pure water. Technically it is defined as "that amount of work that must be done per unit quantity of pure water in order to transport reversibly and isothermically an infinitesimal quantity of water from a pool of pure water, at a specified elevation and at atmospheric pressure to the soil water (at the point under consideration)." This total potential consists of the following potentials.

1. *Matric potential:* That portion of the total soil water potential due to the attractive forces between water and soil solids as represented through adsorption and capillarity. It will always be negative.

2. *Osmotic potential:* That portion of the total soil water potential due to the presence of solutes in soil water. It will generally be negative.

3. *Gravitational potential:* That portion of the total soil water potential due to differences in elevation of the reference pool of pure water and that of the soil water. Since the soil water elevation is usually chosen to be higher than that of the reference pool, the gravitational potential is usually positive.

Solid waste: Waste other than sewage, mostly solid. Household garbage is a familiar form of solid waste.

Solum (plural **sola**): The upper and most weathered part of the soil profile; the A, E, and B horizons.

Solute: A material dissolved in a solvent to form a solution.

Specific gravity: The ratio of the weight or mass of a given volume of a substance to that of an equal volume of a reference substance. Water is the reference for liquids and solids.

Specific surface: The solid particle surface area per unit mass or volume of the solid particles.

Spectral signature: In remote sensing, an objects spectral reflectance pattern. The wavelength and intensity of reflectance.

Splash erosion: The detachment and airborne movement of small soil particles by raindrop impact.

Spoil (overburden): Material overlying sought after resources that is removed and piled to gain access to the minerals, coal, or other underground materials of value. Also, the material remaining after rock has been crushed to remove valuable minerals.

Starter solutions: The term refers to solutions of fertilizers, generally consisting of $N-P_2O_5-K_2O$ in the ratio of 1:2:1 and 1:1:2. These solutions are applied to young vegetable plants at the time of transplanting to help the plants to establish.

Stele: The central conductive tissue of a root or some kinds of stem.

Stereo pair: Two photographs of the same object taken from slightly different positions that when viewed together give the viewer a three dimensional image.

Stomates (or stomata): The controllable openings in the epidermis of leaves and other parts of a plant shoot.

Straight fertilizers: Are the fertilizers containing only one plant nutrient *e.g.* ammonium sulphate, superphosphate, muriate of potash *etc.*

Stratified: Arranged in or composed of strata or layers.

Stress: A force exerted on a body, especially one that strains or deforms its shape. An adverse condition imposed on an organism.

Strip cropping: The practice of growing crops that require different types of tillage, such as row and sod, in alternate strips along contours or across the prevailing direction of wind.

Strip mining: A process in which rock and top soil strata overlying ore or fuel deposits are scraped away to expose the desired deposits.

Stubble mulch: The stubble of crops or crop residues left essentially in place on the land as a surface cover before and during the preparation of the seedbed and at least partly during the growing of a succeeding crop.

Subangular blocky: A structure type whose aggregates are nearly equidimensional, square, or blocklike with rounded edges.

Subsoil: That part of the soil below the plow layer.

Subsoiling: Breaking of compact subsoils, without inverting them, with a special knife-like instrument (chisel), which is pulled through the soil at depths usually of 30-60 cm and at spacings usually of 1-2 m.

Substrata: : Layers of material underlying the soil horizons – for example, sediment or rock.

Substrate: A substratum. More often, the reactant or starting compound for a biochemical reaction or a material that provides food for microbes.

Subsurface diagnostic horizons: One or more soil horizons below the surface horizon with specific properties as described in the soil taxonomy (*e.g.* the argillic horizon).

Subtractive trial: A fertility trial in which deficiencies are detected by observing plant response to the elimination of individual elements from a complete mixture.

Sugar: A simple carbohydrate, consisting of one or two monomer units, each usually containing five or six Catoms, Rapidly used food sources for microbes in soil. Examples are glucose, fructose, and sucrose.

Sulphur coated urea: Urea granules or prills uniformly coated with a layer of sulphur used as slow release fertilizer. The average composition of sulphur coated urea is:

Urea	80-85 per cent
Sulphur	13-16 per cent
Wax	2 per cent

Rate of release of nitrogen from sulphur coated urea is governed by thickness of coating, placement of fertilizer, use of micobicide like coaltar oil, soil temperature, redox potential *etc.*

Summer fallow: A cropping system that involves management of uncropped land during the summer to control weeds and store moisture in the soil for the growth of a later crop.

Superphosphate: This term is most commonly used to denote ordinary superphosphate which contains nearly 16 percent available phosphoric acid. This fertiliseris made by treating rock phosphate with concentrated sulphuric acid.

Surface soil: The uppermost part of the soil, ordinarily moved in tillage, or its equivalent in uncultivated soils. Ranges in depth from 7 to 25 cm. Frequently designated as the "plow layer" the "Ap layer", or the "Ap horizon".

Surface chelation: A chelation reaction that holds a cation at the surface.

Surface tension: The force required per unit length to separate or pull apart a liquid surface.

Suspension: A fluid, usually liquid, bearing small particles held suspended but not dissolved.

Sustainable agriculture: According to Consultative Group on International Agricultural Research (CGIAR), it is the successful management of resources to satisfy the changing needs, while maintaining or enhancing the quality of environment and conserving natural resources.

Symbiosis: The living together in intimate association of two dissimilar organisms, the cohabitation being mutually beneficial.

Synthesis: Combination of simple molecules to form another substance *e.g.*, the union of carbon dioxide and waer under the action of sunlight in photosynthesis. Adjective from synthesis is synthetic, *e.g.* synthetic ammonia.

Talus: Fragments of rock and other soil material accumulated by gravity at the foot of cliffs or steep slopes.

Taxonomy, soil: The science of classification of soils; laws and principles governing the classifying the soil.

Technical classification: A soil classification of inferred rather than observed properties.

Temperature gradient: The rate at which temperature changes with distance, in soil, usually vertical distance.

Tensile strength: The resistance of a soil to forces pulling from opposite directions.

Tensiometer: A device for measuring the negative pressure (or tension) of water in soil in situ; a porous, permeable ceramic cup connected through a tube to a manometer or vacuum guage.

Terrace: 1) A level, usually narrow, plain bordering a river, lake, or the sea. Rivers sometimes are bordered by terraces at different levels. 2) A raised, more or less level or horizontal strip of earth usually constructed on or nearly on a contour and designed to make the land suitable for tillage and to prevent accelerated erosion by diverting water from undesirable channels of concentration; sometimes called diversion terrace.

Tetrahedron: A four sided arrangement of atoms with a cation, such as Si, in the center surrounded by four oxygen atoms or hydroxyls.

Texture: The relative proportion of the various soil separates – sand, silt, and clay – that make up the soil texture classes as described by the textural triangle.

Thermal analysis (differential thermal analysis): A method of analyzing a soil sample for constituents, based on a differential rate of heating of the unknown and standard samples when a uniform source of heat is applied.

Thermophilic organisms: Organisms that grow readily at temperatures above 45°C.

Thiobacillus: Sulphur oxidizing chemoautotrophic bacteria that convert inorganic sulphur to sulphate.

Tile, drain: Pipe made of burned clay, concrete, or ceramic material, in short lengths, usually laid with open joints to collect and carry excess water from the soil.

Till: 1) Unstratified glacial drift deposited directly by the ice and consisting of clay, sand, gravel, and boulders intermingled in any proportion. 2) To plow and prepare for seeding; to seed or cultivate the soil.

Tillage: The mechanical manipulation of soil for any purpose; but in agriculture it is usually restricted to the modifying of soil conditions for crop production.

Tillage, conservation: Any tillage sequence that reduces loss of soil or water relative to conventional tillage, including the following systems.

1. *Minimum tillage*: The minimum soil manipulation necessary for crop production or meeting tillage requirements under the existing soil and climatic conditions.

2. *Mulch tillage*: Tillage or preparation of the soil in such a way that plant residues or other materials are left to cover the surface; also called *mulch farming, trash farming, stubble mulch tillage, plowless farming*.

3. *No-tillage system:* A procedure whereby a crop is planted directly into a seedbed not tilled since harvest of the previous crop; also zero tillage.

4. *Plow-planting*: The plowing and planting of land in a single trip over the field by drawing both plowing and planting tools with the same power source.

5. *Ridge till*: Planting on ridges formed by cultivation during the previous growing period.

6. *Sod planting*: A method of planting in sod with little or no tillage.

7. *Strip till*: Planting is done ina narrow strip that has been tilled and mixed, leaving the remainder of the soil surface undisturbed.

8. *Subsurface tillage*: Tillage with a special sweep like plow or blade that is drawn beneath the surface, cutting plant roots and loosening the soil without inverting it or without incorporating residues of the surface cover.

9. *Wheel track planting*: A practice of planting in which the seed is planted in tracks formed by wheels rolling immediately ahead of the planter.

Tillage, conventional: The combined primary and secondary tillage operations normally performed in preparing a seedbed for a given crop grown in a given geographic area.

Tillage, primary: Tillage that contributes to the major soil manipulation, commonly with a plow.

Tillage, rotary: An operation using a power-driven rotary tillage tool to loosen and mix soil.

Tillage secondary: Any tillage operations following primary tillage designed to prepare a satisfactory seedbed for planting.

Tilth: The physical condition of soil as related to its ease of tillage, fitness as a seedbed, and its impedance to seedling emergence and root penetration.

Top dressing: An application of fertilizer to a soil after the crop stand has been established.

Toposequence: A sequence of related soils that differ, one from the other, primarily because of *topography* as a soil formation factor.

Topsoil: 1) The layer of soil moved in cultivation. 2) Presumably fertile soil material used to top-dress roadbanks, gardens, and lawns.

Tortuosity: A term describing the winding, twisting, and crooked nature of soil pores. A factor contributing to the rate of water movement in soil.

Trace elements: An old term used for the nutrients required in small quantities. Such nutrients are called 'micronutrients' or 'minor elements'.

Transitional horizon: A horizon with properties intermediate between the horizons above and below.

Transitional slide: A break causing a mass to slide along the face of the supporting mass under the force of gravity.

Transmitted energy: In remote sensing, energy that passes through an object.

Transpiration: Evaporation from leaves, the flow of water through plants from soil to atmosphere.

Transported parent material: The opposite of *residual parent material*, unconsolidated material carried by wind or water that, once stabilized, acts as the starting material for soil.

Trial: A term often used to describe a field or container experiment designed to determine nutrient deficiencies.

Triple superphosphate: A fertilizer which contains about 45 to 46 per cent available phosphoric acid. This differs from ordinary superphosphate in that it contains very little calcium sulphate. In the fertilizer trade, the product is also called treble superphosphate or multiple super-phosphate.

Truncated: Having lost all or part of the upper soil horizon or horizons.

Tuff: Volcanic ash usually more or less stratified and in various states of consolidation.

Tundra: A level or undulating treeless plain characteristic of arctic regions.

Turgulent transfer: Diffusion-like movement of heat, gases, and solutes greatly accelerated by irregular mixing motion – turbulence – in the fluid medium.

Turgor: The normal state of turgidity in living cells produced by pressure of the cell contents on the cell walls.

Turgor pressure: Pressure exerted on a cell wall by the cell's constituents (and vice versa).

Ultisols: Strongly weathered soils formed in warm, humid regions under forest vegetation.

Undifferentiated group: A soil map unit consisting of two or more similar soil units not in a regular geographic association. The soil units have the same or very similar use and management.

Unit cell: The smallest deinable repeating structural unit in a crystalline material.

Universal soil loss equation (USLE): An equation for predicting the average annual soil loss per unit area per year, A = RKLSPC, where R is the climatic erosivity factor (rainfall plus runoff), K is the soil erodibility factor, L is the length of slope, S is the percent slope, C is the cropping and management factor, and P is the soil erosion practice factor.

Unsaturated flow: The movement of water in a soil that is not filled to capacity with water.

Urease: A crystalizabale protein enzyme that activates hydrolysis of urea.

$$NH_2CO.NH_2 + 2H_2O \longrightarrow (NH_4)^2CO_3$$

Urea   Water   Amm.Carbonate

Its richest source is soyabean. It is found extensively in jack beans and numerous fungi. It is marketed as white tablets or powder. It is soluble in dilute alkali. Its isoelectric point is pH 5.5. It is activated by the presence of metals. Urease is found in soils in various degrees. If its concentration is high, urea hydrolysis in soils takes place very rapidly. Urease enzyme taken from soyabean or jack bean is also used to determine urea in mixed fertilizers.

Uronite: A sugar with a COOH group.

Van der Waal's forces: Combined polar and London attractions between molecules in close proximity.

Variable charge: An electrical charge on clay or organic matter that changes with changes in soil pH.

Varnish, desert: A glossy sheen or coating on stones and gravel in arid regions.

Vascular: Having to do with veins.

Vermiculite: A 2: 1-type silicate clay usually formed from mica that has a high net negative charge stemming mostly from extensive isomorphous substitution of aluminum for silicon in the tetrahedral sheet.

Vertical photo: In remote sensing, a photo taken from directly above an object rather than at an angle to the object.

Vertisols: Soils high in clay that form deep cracks at least one centimeter wide during dry season.

Vesicular arbuscular mycorrhiza: A common endomycorrhizal association produced by phycomycetous fungi of the genus *Endogone* and characterized by the development of two types of fungal structures: a) within root cells small structures known as arbuscles and b) between root cells storage organs known as vesicles. Host range includes many agricultural and horticultural crops.

Vesicular pore: A soil pore not connected to other pores.

Virgin soil: A soil that has not been significantly disturbed from its natural environment.

Visible spectrum: Electromagnetic radiations which are within the range of 400-700 nm and visible to the naked eye. The beam formed due to these is called the visible spectrum. It consists of seven colours corresponding to the following wavelength:

Colour of Light	Range of Wavelength (in nm)
Red	747-700
Orange	585-647
Yellow	575-585
Green	491-575
Blue	424-491
Violet	400-424

Void ratio: A measurement of porosity. The ratio of void volume to soil bulk volume.

Volcanic ash: Fine particles of rock blown into the air by a volcano. Settles, often in layers, to become soil parent material. Often consists of short-range-order minerals.

Volumetric analysis: A group of methods of quantitative analysis based on accurate measurement of the volume of a reagent of known concentration used in the analysis. These are extensively used in fertilizer analysis of three types: Neutralization, oxidation reduction and complex formation.

Washed in zone: Formed by clay separating from the top few tenths of a millimeter of soil and being translocated into the next few tenths to clog the pores.

Water erosion: The natural wearing away of the earth's surface by rainfall and surface runoff.

Waterlogged: Soils saturated with water.

Water potential: The tendency of soil water to move, the sum of gravity, pressure, matric and solute components.

Water profile: The pattern of vertical variation in water content or potential with depth in the soil.

Water retention curve or soil water characteristic curve: The graphical relationship between soil water content (by mass or colume) and the soil water matric potential (the energy required to remove the water).

Water stable aggregate: A soil aggregate stable to the action of water such as falling drops, or agitation as in wet-sieving analysis.

Water table: The upper surface of groundwater or that level below which the soil is saturated with water.

Water table, perched: The surface of a local zone of saturation held above the main body of groundwater by an impermeable layer of stratum, usually clay, and separated from the main body of groundwater by an unsaturated zone.

Water use efficiency: Dry matter or harvested portion of crop produced per unit of water consumed.

Weathering: All physical and chemical changes produced in rocks, at or near the Earth's surface, by atmospheric agents.

Wetting front: The advancing boundary between dry soil and wetted soil during infiltration.

Wilting point (permanent wilting point): The moisture content of soil, on an oven-dry basis, at which plants wilt and fail to recover their turgidity when placed in a dark humid atmosphere.

Windbreak: Planting of trees, shrubs, or other vegetation perpendicular, or nearly so to the principal wind direction to protect soils, crops, homesteads, *etc.*, from wind and snow.

X-ray diffraction: An analytical method that involves the exposing of samples to X-rays. The radiation is reflected from the sample in response to structure of the crystals that compose the sample. It is useful in identification of elements in materials and coatings.

Xeric: Refers to a dry environment *e.g.* desert.

Xerophytes: Plants that grow in or on extremely dry soils or soil materials.

Xenon: A nobal gas, which found in atmosphere in trace amount.

Xylem: One of two main kinds of conductive tissue in plants. Xylem has tubelike cells that conduct water and ions rapidly from roots to leaves. Wood is xylem tissue.

Yellow cake: Uranium oxide (U_3O_8) that results from the refining of uranium ore. The purified material contain 93.3 percent uranium-238 and 0.7 percent uranium-235. The term yellow cake is applied because of the colour and texture of the material.

Yield: The individuals or biomass removed when the plant population is harvested.

Yield potential: The total production capacity of a crop.

Zero order reaction: A chemical reaction in which the rate of reaction is independent of reactants concentration.

Zero tillage: It is a system which refers to planting crops with minimum of soil disturbance. In this seed are placed directly into narrow slits 2 – 4 cm wide and 4 – 7 cm deep made with a drill fitted with chisel "inverted T" or double disc openers without land preparation. In this soil macro and micro fauna builds and maintain an open-pore structure of the soil.

Zeolite: Natural hydrated silicate of calcium and aluminium used in water softening process by the ion exchange method.

Zone of depletion: Top layer of A horizon of a soil, characterised by the removal of soluble and insoluble soil components through leaching and eluviation by gravitational water.

Zinc (Zn): An essential plant micronutrient found deficient in a variety of cultivated soils all over the world. Its atomic number in 30 and atomic weight is 65.37. Total zinc content in soils varies from 5 – 50 ppm. In some soils its deficiency is attributed to phosphate application. Plants contain zinc in the range of 1-50 ppm.

Index

www.ingramcontent.com/pod-product-compliance
Lightning Source LLC
Chambersburg PA
CBHW031953180326
41458CB00006B/1708